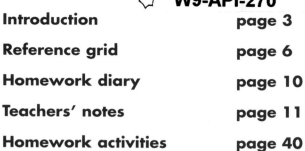

W9-API-270

Introduction	page 3
Reference grid	page 6
Homework diary	page 10
Teachers' notes	page 11
Homework activities	page 40

100 SCIENCE HOMEWORK ACTIVITIES

Published by
Scholastic Ltd,
Villiers House,
Clarendon Avenue,
Leamington Spa,
Warwickshire CV32 5PR

Printed by Ebenezer Baylis and Son Ltd, Worcester

© Scholastic Ltd 2002
Text © 2002 Clifford Hibbard, Karen
Mallinson-Yates, Ian Mitchell and Tom Rugg

1 2 3 4 5 6 7 8 9 0   2 3 4 5 6 7 8 9 0 1

Authors
Tom Rugg (Year 5 Unit 1, Year 6 Units 1–3),
Ian Mitchell (Year 5 Units 2–3),
Karen Mallinson-Yates (Year 5 Units 4 & 8, Year 6 Units 4 & 8),
Clifford Hibbard (Year 5 Units 5–7, Year 6 Units 5–7)

Editor
David Sandford

Assistant Editor
Clare Gallaher

Series Designer
Micky Pledge

Designer
Rachel Warner

Cover photography
Martyn Chillmaid/Photodisc

Illustrations
Beverly Curl

British Library Cataloguing-in-Publication Data
A catalogue record for this book is available from the
British Library.

ISBN 0-590-53725-3

The rights of Tom Rugg (Year 5 Unit 1, Year 6 Units 1–3), Ian
Mitchell (Year 5 Units 2–3), Karen Mallinson-Yates (Year 5
Units 4 & 8, Year 6 Units 4 & 8) and Clifford Hibbard (Year 5
Units 5–7, Year 6 Units 5–7) to be identified as the Authors of
this work have been asserted by them in accordance with the
Copyright, Designs and Patents Act 1988.

Teachers should consult their own school policies and
guidelines concerning practical work and participation of
children in science experiments. You should only select
activities which you feel can be carried out safely and
confidently in the classroom.

Acknowledgements

The National Curriculum for England 2000
© The Queens Printer and Controller of HMSO.
Reproduced under the terms of HMSO Guidance
Note 8.

A Scheme of Work for Key Stages 1 and 2: Science
© Qualifications and Curriculum Authority.
Reproduced under the terms of HMSO Guidance
Note 8.

Data for 'Endangered species in our skies' on page
54 taken from the British Trust for Ornithology
(www.bto.org)

Data for 'Has the cod had its chips?' on page 62
from Centre for Environment, Fisheries and
Agriculture (www.cefas.gov.uk) © Crown copyright.

Every effort has been made to trace copyright holders
and the publishers apologise for any omissions.

The publisher has made every effort to ensure that
websites and addresses referred to in this book are
correct and educationally sound. They are believed
to be correct at the time of publication. The
publishers cannot be held responsible for subsequent
changes in the address of a website, nor for the
content of the sites mentioned. Referral to a website
is not an endorsement by the publisher of that site.

100 SCIENCE HOMEWORK ACTIVITIES

100 Science Homework Activities is a series of resource books for teachers of Years 1–6 (Scottish Primary 1–7). Each book of 100 activities covers two year groups, with around 50 activities specific to each year. These provide a 'core' of homework tasks in line with the National Curriculum documents for science in England, Wales and Northern Ireland and, in England, the QCA's *Science Scheme of Work*. The tasks also meet the requirements of the 5–14 National Guidelines for science in Scotland.

The homework activities are intended as a support for all science teachers, be they school science leader or trainee teacher. They can be used with any science scheme of work as the basis for planning homework activities throughout the school in line with your homework policy. If you are using the companion series, *100 Science Lessons*, these books are designed to complement the lesson plans in the corresponding year's book. Activities can be used with single- or mixed-age classes, single- and mixed-ability groups, and for team planning of homework across a year or key stage. You may also find the activities valuable for extension work in class, or as additional resources.

Using the books

100 Science Homework Activities has been planned to offer a range of simple science exercises for children to carry out at home. Many are designed for sharing with a helper, who could be a parent or carer, another adult in the family, an older sibling, or a neighbour. They include a variety of games, puzzles, observations and practical investigations, each of which has been chosen to ensure complete coverage of all UK national curricula for science.

Teacher support

There are supporting teachers' notes for each of the 100 activities in this book, briefly outlining the following:

Learning objectives: the specific learning objectives that the homework aims to address, based on the four curriculum documents, and linked to the same learning objectives from the relevant *100 Science Lessons* book.

Lesson context: a brief description of the classroom experience recommended for the children before undertaking the homework activity.

Setting the homework: advice on how to explain the worksheet to the children, and how to set it in context before it is taken home.

Back at school: suggestions for how to respond to the completed homework, including discussion with the children or specific advice on marking, as well as answers where relevant.

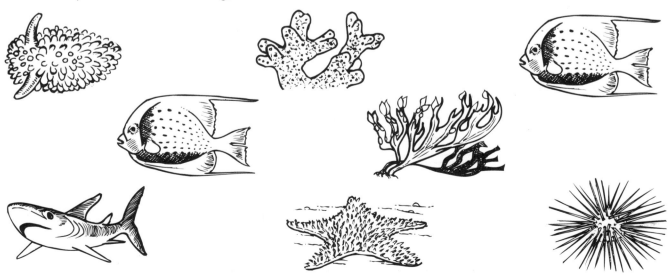

Photocopiable pages

Each of the 100 homework activities in this book includes a photocopiable worksheet for children to take home. The page provides instructions for the child and a brief explanation of the task for the helper, stating simply and clearly the activity's purpose and suggesting ideas for support or a further challenge to offer the child. The science topic addressed by each activity, and the type of homework being offered, are both indicated at the top of each page. There are seven types of homework activity:

Science to share activities encourage the child, with their helper, to talk and work together on a science task. These tasks draw heavily on things likely to be found at home.

Science practice/revision activities are tasks designed to reinforce knowledge or understanding gained during lesson time.

Numeracy/Literacy link activities practise skills from other areas of the curriculum within a science context.

Finding out activities are designed to increase children's knowledge through investigations, keeping diaries, or by consulting simple secondary sources.

Observation tasks require children to look closely and carefully at things around the home to gain more detailed knowledge of a science topic.

Ask an adult activities help children to understand that asking questions is a valuable way of finding out more about a particular subject, particularly when they are too young to have experienced a particular activity or event themselves.

The grids on pages 6–9 provide an overview of the book's content, showing how each activity can be matched to the curriculum: in England to the National Curriculum for science and the QCA *Science Scheme of Work*, and in Scotland to the 5–14 National Guidelines.

Using these activities with the *100 Science Lessons* series

The organisation of the homework activities in this book matches that of the activities in *100 Science Lessons: Year 5* (written by David Glover, Ian Mitchell, Louise Petheram and Peter Riley) and *100 Science Lessons: Year 6* (written by Clifford Hibbard, Karen Mallinson-Yates and Tom Rugg), both published by Scholastic, so that there are homework activities matching the learning objectives in each unit of work. The grids on pages 6–9 in this book show which lessons in *100 Science Lessons: Year 5* and *Year 6* have associated homework activities here, together with the relevant page numbers from the main books to help with planning.

Supporting your helpers

As well as the notes on each of the worksheets, there is a photocopiable homework diary – provided on page 10 – which can be sent home with each of the homework activities. The diary has space for recording four pieces of homework, and multiple copies can be stapled together to make a longer-term homework record. There is space to record the activity's title and the date it was sent home, and spaces for responses to the homework from the child, the helper and your own comment. The diary is intended to encourage home–school links, so that parents or carers know what is being taught and can make informed comments about their child's progress.

PHOTOCOPIABLE HOMEWORK DIARY

Homework diary   Name:_____

Name of activity	Date sent home	Did you like this? / What did you learn? Write down what you thought about the homework.		Helper's comments	Teacher's comments
Where has the song thrush gone?	2nd May	I'm not really very interested in birds, but I'm trying to spot a song thrush.	I found out that slug pellets might harm the song thrush. I've told my gran not to use them!	We used the Internet together to find information about this bird. Interesting.	Glad you shared the Internet together. It's interesting to see Gran's having her habits changed!

About this book: Years 5&6/Primary 6–7

This book provides 100 creative science homeworks for nine- to eleven-year-olds. Each unit offers between three and nine activities to support that topic, so even if you choose not to use all the activities for homework there is plenty of choice, and a wealth of suitable extension material for use in the classroom, particularly with classroom helpers.

Older primary-aged children need to make sense of their science learning by seeing, applying and relating it to experiences in their everyday lives. Although there are a variety of activities in this book, children's learning will be enhanced in every case by the chance to share it with a helper. The majority of the activities need no special resources, using only simple household equipment or things that are commonly available at home. Many encourage the children to make use of their local environment, or to draw on experience of their surroundings, where adult supervision is clearly essential.

Progression

These activities for Years 5 and 6/Scottish Primary 6–7 cover all areas of the National Curriculum and, to meet Scottish requirements, activities on energy and the Solar System, for example, have been included. There is clearly progression between the activities for Year 5 and Year 6, and those in Year 6 aim to prepare children for the Key Stage 2 National Tests through revision and practice activities.

The activities build on science work in earlier years, and aim to extend the children's work by asking them to think about their own lives and to apply this knowledge to what they have learned in the classroom. The follow-up classwork suggestions are designed to encourage the pupils to take a more active part in their own learning.

Science coverage

Many of the activities in the 'Environment' units (Units 2 and 3 in each year's activities) provide opportunities for education relating to sustainable development – considering water and electricity consumption, conserving fish stocks, encouraging biodiversity in the garden, and aim to encourage children to take a more active role in their own learning. There is clear research evidence, published by the Centre for Sustainable Energy (tel 0117 929 9950; www.cse.org.uk), indicating that older primary-aged children can be incredibly effective at promoting less consumptive and more sustainable home lifestyles.

Homework activities for 'Materials and their properties' allow the children to explore the ways in which some everyday materials, such as water and chocolate, change when they are heated and cooled, and how some changes are reversible and others are non-reversible. The activities also allow the children to apply their knowledge of separating mixtures or materials, and to develop their awareness of health and safety issues and their investigation skills.

Physical processes work in Year 5/Primary 6 may provide opportunities to link to science and technology, for example through QCA Unit 5/6H. The homework activities seek to take this work into the home context. By this stage, children should be able to relate some of what they are learning to the real world, so there are a number of activities in simplified 'real life' contexts: the science of the theme park, or who wears glasses and why, for example. Activities on environmentally friendly power supply could make useful links to work in Units 2 and 3.

The physical processes activities in Year 6/Primary 7 inevitably seek to practise skills that are important for children in England facing national tests, but try to do so in ways less obvious than the usual 'quiz'. To this end you will find lots of work on interpreting graphs, for example, which has been shown consistently to cause children taking tests difficulty. There are also some more sophisticated 'real life' contexts, such as how a bank keeps money safe, and the forces involved in extreme sports. Space (always a hot topic with at least one 'class expert'!) reaches out to the wider Solar System, as this is a key topic in the Scottish 5–14 Guidelines and usually of interest to children everywhere.

YEAR 5 REFERENCE GRID

Page in this book	Activity name	Homework type	Learning objectives	QCA Unit	National Curriculum	Scottish 5–14 Guidelines	Unit	Lesson	Page
40	My food and drink diary	Science to share	To be able to group foods according to their relative nutritional value; To interpret data in a table	5A	Sc2 1a, b; 2a	Taking responsibility for health	1	1	13
41	What is digestion?	Literacy link	To know that food is digested by organs, together called the digestive system	5A	Sc2 1a	Identify the organs of the body – C; Outline the process of digestion – D	1	4	15
42	Gills and lungs	Science practice	To know the basic structure of the respiratory system	5A	Sc1 1a, b; 2g, i; 2d, h	Make predictions – C; Make an accurate set of measurements – D	1	5	17
44	Does heart rate decrease with age?	Science to share	To know the pulse is produced by the heart beat; To carry out an investigation, make observations and draw conclusions	5A	Sc2 2c, d, h	Make predictions – C; Make an accurate set of measurements – D	1	6–7	19–21
45	Changes at puberty	Science practice	To know about the changes which take place in the body at puberty	5A	Sc2 1a, 2f	Describe the changes that occur during puberty – D	1	9–10	22–3
46	Smoking	Numeracy link	To know how tobacco affects the body	5A	Sc2 2g	Taking responsibility for health	1	12	27
47	The life cycle of a bean	Science to share	To know that flowering plants have life cycles; To know the factors that affect germination; To develop observation skills	5B	Sc1 1b; Sc2 3a, b, c, d;	Describe the main stages of plant reproduction – D	2	1–9	43–54
49	Tree seed detective	Science practice	To know that seeds must be dispersed to help new plants grow	5B	Sc2 3d	Describe the main stages of plant reproduction – D	2	3	45
50	Insects and flowers	Science practice	To know the function of pollen, the stamens and the stigma, that seeds develop in the ovary and the role of insects in pollination	5B	Sc2 3d	The processes of life – D	2	9	52
51	Frogs in the garden	Literacy link	To develop an understanding of life cycles; To know the life cycle of a frog	5B	Sc2 1a	The processes of life – B; Interaction of living things with their environment – B, C	2	10	54
53	Where has the song thrush gone?	Finding out	To understand that if living things fail to reproduce they become extinct	5B	Sc2 5a	Interaction of living things with their environment – C, D	2	12	58
54	Endangered species in our skies	Numeracy link	To understand that if living things fail to reproduce they become extinct	5B	Sc2 5a	Interaction of living things with their environment – C, D	2	12	58
55	The life cycle of a border collie	Science practice	To know that mammals have life cycles	5B	Sc2 1a, c	The processes of life – C, D	2	13–15	60–3
56	The water cycle in the kitchen	Science practice	To know that water circulates through the environment	5B	Sc3 1e, 2d, e	Materials from Earth – C, D; Changing materials – C	3	1–2	75–7
57	Getting rid of water	Finding out	To recognise things in our home that remove water	–	Sc1 2h	Describe human impact on the environment – D	3	4–5	78–80
58	Saving water saves money	Literacy link	To know that we use a great deal of water and that water must be conserved	5H	Sc2 5a	Materials from Earth – B	3	6	80
60	A diary of a mini-pond	Science to share	To create a habitat with simple safe equipment; To make systematic observation and recording	5H	Sc1 2e, f	Investigating – C	3	7–8	82–3
61	Reducing pollution	Science practice	To recognise activities that may cause air pollution, their effects on the environment and how they can be reduced	5H	Sc2 5a	Interaction of living things with their environment – C, D	3	9	84
62	Has the cod had its chips?	Numeracy link	To understand that an indigenous fish species is in danger of extinction and needs to be conserved if it is to survive	–	Sc2 5a	Interaction of living things with their environment – C, D	3	10	86
63	Gas to liquid	Science practice	To reinforce that some materials can be classified as solids, and some as liquids; To know that air is a gas and has mass	5C, 5D	Sc3 1e, 2b, d	Describe the differences between solids, liquids and gases – C	4	1	99
64	Air	Finding out	To know that air is all around us; To know that air is a type of material called a gas	5C, 5D	Sc3 1e	Describe the differences between solids, liquids and gases – C	4	2	99
65	Solid, liquid or gas?	Science to share	To know that gases do not have a fixed volume and can be compressed; To know that gases and liquids can flow	5C, 5D	Sc3 1e	Describe the differences between solids, liquids and gases – C	4	8	105
66	Steamy	Science practice	To know that gases can be turned into liquids by cooling; To know water vapour is present in air and cannot be seen	5C, 5D	Sc3 2b, d	Describe in simple terms the changes when water is heated or cooled – C	4	9	107
67	Liquid to gas	Numeracy link	To know that when a liquid is heated to a certain temperature, it boils and turns into a gas; To know that water boils at 100°C	5C, 5D	Sc1 2h; Sc3 2b, d, e	Describe in simple terms the changes when water is heated or cooled – C	4	10	109
68	What a state!	Literacy link	To know that heating/cooling can cause materials to melt, boil, evaporate, freeze or condense	5C, 5D	Sc3 2b, d	Describe in simple terms the changes when water is heated or cooled – C	4	12	112

Page in this book	Activity name	Homework type	Learning objectives	QCA Unit	National Curriculum	Scottish 5–14 Guidelines	Unit	Lesson	Page
69	Solid to liquid	Science practice	To conduct an investigation into rates of melting; To know that the freezing/melting temperature of water is 0°C	5C, 5D	Sc3 2e	Describe in simple terms the changes when water is heated or cooled – C	4	13	113
70	Going on and on and on...	Numeracy link	To know that in a battery, different chemicals are brought together to produce electricity	–	Sc4 1a	–	5	1	120
71	Locked out	Science practice	To know how an electromagnet can be constructed and used	–	Sc4 1a, 2a	Provide reasons for planning decisions – D	5	2	121
72	Spinning around	Science to share	To know some uses of electric motors	–	Sc4 1a, 2a	Give examples of energy being converted – C	5	4	124
73	Simon's new crane	Science practice	To use electric motors in toys; To understand circuit drawings including motors	–	Sc4 1c	Give examples of energy being converted – C	5	5	126
74	We have the power	Finding out	To realise that power generation has environmental implications	–	Sc2 5a; Sc4 1a, 2a	–	5	6	126
75	Emission control	Literacy link	To know that 'environmentally friendly' power sources exist	–	Sc4 1a	Give examples of energy conversions involved in generating electricity – D	5	7	127
76	Mind the gap	Finding out	To know that a variety of structures are used to support the weight of objects	5/6H	Sc1 1a; Sc4 2b, d, e	–	6	1	135
77	A day out at Adventure Island	Literacy link	To revise that friction can be reduced by lubricants	5/6H	Sc1 1a, 2i; Sc4 2c	Provide reasons for planning decisions – D	6	3	137
78	Loosen up	Ask an adult	To know that friction can be reduced by lubricants	5/6H	Sc4 2c	–	6	4	138
79	Ball-bearing clocks	Science practice	To know that everything falls down due to gravity; To know that falling materials can be used to measure time	5/6H	Sc1 2i, j; Sc4 2b	Give examples of energy being converted – C	6	5	139
80	Sail racing	Numeracy link	To know that the force of the wind can be used to provide energy for work	5/6H	Sc1 2e; Sc4 2e	Give examples of energy being converted – C	6	10	144
81	A sudden jump	Science practice	To know that energy stored in a spring band can be turned into movement energy	5/6H	Sc1 2e, Sc4 2e	Give examples of energy being converted – C	6	11	145
82	Crooked pencils	Science to share	To know that when light passes through curved, transparent material the paths of the rays are changed	5F	Sc4 3a	Describe what happens when light passes through different materials – E	7	1	154
83	Who wears the glasses?	Numeracy link	To know that the eye contains a lens and a place where the image forms	5F	Sc4 3d	Describe how lenses work – C	7	2–4	155–7
84	Primary colours	Finding out	To know that sunlight is made from a range of different-coloured light rays	5F	Sc1 2j; Sc4 3d	Explain the effect of a prism on white light – E	7	6	159
85	Ding dong	Science practice	To know that vibrating objects produce different sounds	5F	Sc1 2j, l Sc4 3e	Link sound to sources of vibration – C	7	7	161
86	I can't hear you!	Numeracy link	To know that sound travels in straight lines; To know that sound can travel through air	5F	Sc4 3g	Explain what happens when sound passes through different materials – E	7	8–9	162–4
87	My music	Literacy link	To know that the pitch and volume of some musical instruments can be altered by changing how they vibrate	5F	Sc4 3e, f, g	Use pitch and volume to describe sound – D	7	12–14	167–70
88	Is the Earth round?	Literacy link	To know that the Sun, Earth and Moon are approximately spherical	5E	Sc1 2b; Sc4 4a	Describe the Solar System in terms of the planets – C	8	1	181
89	Day and night	Science practice	To use knowledge to explain observations; To know that the apparent movement of the Sun is caused by the Earth's rotation	5E	Sc1 2h; Sc4 4b, c, d	–	8	3	183
90	Daylight and darkness	Observation	To use knowledge to explain observations; To know that the apparent movement of the Sun is caused by the Earth's rotation	5E	Sc1 2h; Sc4 4b, c, d	–	8	3	183
91	Making a sundial	Science to share	To know that time varies around the world; To see how some units of time are linked to the Earth's motion but others are artificial	5E	Sc4 4c	Explain day, month and year in terms of the movement of the Earth – E	8	4	186
92	Day-lengths	Numeracy link	To know that other planets have different day-lengths from Earth; To think logically about causes and effects	5E	Sc1 1a	–	8	5	186
93	The four seasons	Science practice	To know that the Earth travels in an orbit around the Sun once a year, and this causes the seasons we experience	5E	Sc4 4d	Explain day, month and year in terms of the movement of the Earth – E	8	7	187

REFERENCE GRID YEAR 5

YEAR 6 REFERENCE GRID

Page in this book	Activity name	Homework type	Learning objectives	QCA Unit	National Curriculum	Scottish 5–14 Guidelines	Unit	Lesson	Page
94	The Earth's 23½° tilt	Literacy link	To know that the Earth travels in an orbit around the Sun once a year, and this causes the seasons we experience	5E	Sc4 4d	Explain day, month and year in terms of the movement of the Earth – E	8	8	189
95	Family fitness plan	Science to share	To construct a plan for a healthy lifestyle	–	Sc1 1a, 2i; Sc2 2b, h	Taking responsibility for health and safety	1	1	12
96	Lifestyle muddle	Science practice	To identify activities and substances that may be harmful to health	–	Sc2 2g	Taking responsibility for health and safety	1	2	14
97	Staying alive!	Numeracy link	To know the stages of the human life cycle; To assess the healthy and unhealthy aspects of a lifestyle	–	Sc2 2f	Describe the main stages in human reproduction – D	1	4	16
98	Piggy-back tadpoles	Literacy link	To know about the skills and care required in parenting	–	Sc2 2f	Respect and care of others	1	6	20
99	Reproduction crossword	Science practice	To know about the structure of the reproductive organs, how fertilisation occurs and how the foetus grows	–	Sc2 2f	Identify and name the main organs of the reproductive system – E	1	6–7	20–2
100	Animal parents	Finding out	To know about the skills and care required in parenting	–	Sc2 2f	Respect and care of others	1	6–7	20–2
101	Coral reefs	Finding out	To know that living things can be arranged into groups based upon observable features	6A	Sc2 4b	Name common vertebrates – C	2	2	34
102	Ancient animals	Science practice	To know that living things can be put into groups based upon observable features and that this can help identify creatures	6A	Sc2 4b	Name common vertebrates – C	2	2	34
103	A guide to my family and friends	Science practice	To use branching keys to identify an organism	6A	Sc2 4a, c	Name common animals and plants using keys – C	2	3	36
105	Variations	Science practice	To know that animals of the same species vary	6A	Sc1 2f, i	Make an appropriate series of accurate measurements – D	2	4	38
106	Plants in space	Literacy link	To know that plants need light, water and warmth to grow well	6A	Sc2 3a	Identify the conditions and products of photosynthesis – E	2	7	42
107	Growing tomatoes	Numeracy link	To know that plants need nutrients from the soil for healthy growth, and that farmers and gardeners often add these as fertiliser	6A	Sc2 3c	–	2	8	42
108	Garden food webs	Science practice	To know that food chains are used to describe feeding relationships in a habitat	6A, 6B	Sc2 5d, e, f	Construct and interpret simple food webs – E	3	2	53
109	Adaptations for survival	Science practice	To know that plants and animals have special features that help them to survive in a habitat	6A, 6B	Sc2 5b, c	Give examples of how plants and animals are suited to their environment – D	3	4	56
110	Lichens and air pollution	Numeracy link	To use simple environmental survey techniques	6A, 6B	Sc1 2f, h; Sc2 5b, c	Select an appropriate way of recording findings – D	3	5	58
111	The soil where I live	Finding out	To know that different soils can be compared	6A, 6B	Sc1 2e, i, j	–	3	7	60
112	Preventing decay	Literacy link	To know that decay caused by micro-organisms is useful in the environment	6A, 6B	Sc2 5f	Give the main distinguishing features of micro-organisms – E	3	10	64
113	Useful microbes	Science practice	To know that micro-organisms are used in food production	6A, 6B	Sc2 5f	Give the main distinguishing features of micro-organisms – E	3	15	70
114	Soluble or insoluble?	Science to share	To know that some materials dissolve in water, and that solids that do not dissolve in a liquid can be separated by filtering	6C	Sc1 2e; Sc3 2a, d; 3c	Changing materials – C, D	4	1	83
115	Ice and water	Science practice	To know that when a gas is cooled it becomes a liquid and that this process is called condensing	6C	Sc3 2d, 2e	Changing materials – C, D; Materials from Earth – C	4	3	88
116	The dissolving sugar experiment 1	Numeracy link	To know how the temperature of water affects the speed of dissolving; To be able to interpret results	6C	Sc3 2d, 3b; Sc1 1a, 2i	Changing materials – E; Reviewing and reporting on tasks – C, D	4	4–5	90–2
117	The dissolving sugar experiment 2	Numeracy link	To know how the temperature of water affects the speed of dissolving; To be able to interpret results	6C	Sc1 2i	Changing materials – E; Reviewing and reporting on tasks – C, D	4	4–5	90–2
118	Mixing materials	Science practice	To understand that changes sometimes happen when materials are mixed, and these changes cannot easily be undone	6C	Sc3 2a, 2f; Sc1 2i	Changing materials – C	4	9	98
119	Burning	Science practice	To know that burning brings about changes that are irreversible	6C	Sc3 2f	Changing materials – D	4	12–13	102–4

Page in this book	Activity name	Homework type	Learning objectives	QCA Unit	National Curriculum	Scottish 5–14 Guidelines	Unit	Lesson	Page
120	Rusting	Literacy link	To know that only iron and steel rust; To know that oxygen and water are needed for rusting and that rusting is irreversible	6C	Sc3 2f	Changing materials – E; Reviewing task: Interpreting processes – D, E	4	15	106
121	Generating electricity	Literacy link	To know that electricity is made from non-renewable fuels	6C	Sc3 2g	Explain the difference between renewable and non-renewable energy – E	4	19	111
122	Traffic lights	Science practice	To know that the number of batteries and bulbs in a circuit can affect the brightness of the bulb(s)	6G	Sc1 1b, 2e; Sc4 1a, b	Describe the effect of changing components in a circuit – D	5	1–2	124–6
123	Matching up	Science practice	To know that electrical circuits and components can be represented by conventional symbols	6G	Sc4 1c	Construct a circuit diagram using symbols – D	5	3	126
124	The combination is...	Science practice	To know that switches can be placed between parts of a circuit to provide alternative routes for current to pass along	6G	Sc4 1a	Construct simple circuits, identifying main components – C	5	4–6	128–9
125	The safety code	Literacy link	To recognise the dangers associated with electricity	6G	–	Taking responsibility for health and safety	5	7	129
126	Wired up	Numeracy link	To know that the amount of electricity flowing in a circuit is related to the total resistance	6G	Sc4 1a	To use *resistance* in the context of a circuit – E	5	8	131
127	Many hands make light work	Science to share	To know about the construction of electrical circuits in real life	6G	Sc4 1c	Construct simple circuits, identifying main components – C	5	10	132
128	Your next game is...	Science practice	To know that a force exists between two magnets, and between magnets and magnetic materials	6E	Sc1 1b, 2a, c; Sc4 2a	–	6	1	142
129	Newton's diary	Literacy link	To know that the force of gravity is responsible for the weight of an object	6E	Sc1 1a, b; Sc4 2b	Describe the effect of gravity on an object – D	6	3	144
130	Bungee!	Numeracy link	To know that when the force stretching an elastic band is increased, its length increases in proportion	6E	Sc1 2h; Sc4 2d	Describe balanced and unbalanced forces – E	6	4	146
131	Floating metal	Science to share	To know that when an object floats, the upthrust acting on it is equal to the force of gravity and acting in the opposite direction	6E	Sc1 2l; Sc4 2e	Describe balanced and unbalanced forces – E	6	6	147
132	The daredevil	Literacy link	To be able to describe how air resistance can slow down a falling (moving) object	6E	Sc4 2c	Describe air resistance in terms of friction – C	6	7–8	148–9
133	The gymnastics display team	Science practice	To be able to explain how objects stay at rest or move by considering the forces acting on them	6E	Sc4 2e	Describe balanced and unbalanced forces – E	6	10	150
134	Staying in the shade	Science practice	To revise how light travelling from a source can be blocked by an opaque object, making a shadow	6F	Sc4 3a, b	Link light to shadow formation – C	7	1	161
135	True or false?	Science practice	To know that non-luminous objects are seen because light scattered from them enters the eye	6F	Sc4 3c	Give examples of light being reflected – C	7	3	162
136	Seeing things differently	Science to share	To know that mirrors can change the direction in which light travels; To know that mirrors are shiny surfaces, not dull	6F	Sc4 3c	Give examples of light being reflected – C	7	5–7	164–7
137	What a view!	Literacy link	To recognise shadows and reflections in the environment, and know the difference	6F	Sc4 3b, c	Link light to shadow formation – C	7	8	167
138	On the way	Science practice	To reinforce the idea that parts vibrate to produce sound; To know how pitch and loudness can be changed	6F	Sc4 3e, f	Use 'pitch' and 'volume' to describe sound – D	7	9	167
139	The speed of sound	Numeracy link	To reinforce that sound can travel through solids, liquids and gases	6F	Sc1 2c; Sc4 3g	Make a report of an investigation – C, D	7	11	169
140	The Sun, Moon and the Earth	Science practice	To know that the Earth spins as it goes around the Sun, and that the Moon travels with the Earth and in an orbit around it	–	Sc 4 4a, b, c, d; Sc1 2i	Earth in space – A, B; Reviewing and reporting on tasks – C, D	8	1	179
141	An eclipse of the Sun	Literacy link	To know how a solar eclipse occurs.	–	Sc 4 4a, b, d	Earth in space – C, E; Reviewing and reporting on tasks – C, D	8	2	181
142	The life and work of Galileo Galilei	Finding out	To be able to describe the surface of the Moon; To know how the craters on the Moon were formed	–	Sc4 4d; Sc1 1a, b, 2d, g h, k, m	Earth in space – C; Preparing for tasks – C, D; Carrying out tasks – C, D	8	3	183
143	The Solar System	Science practice	To know what makes up our Solar System; To know the order of the planets, and which planets have moons	–	–	Earth in space – C, D	8	4	184
144	How far from the Sun?	Numeracy link	To know the order of the planets in our Solar System	–	–	Earth in space – C, D	8	4	184

Homework diary

Name: _____

Name of activity	Date sent home	Did you like this? What did you learn? Write down what you thought about the homework.		Helper's comments	Teacher's comments

Teachers' notes

UNIT 1 OURSELVES | GROWING UP HEALTHY

p40 My food and drink diary SCIENCE TO SHARE

Learning objectives
- To be able to group foods according to their relative nutritional value.
- To use a table to record and interpret data.

Lesson context
Introduce or revise the five different food groups (proteins, fats, carbohydrates, fibre, vitamins and minerals) and their relative importance in contributing to a healthy, balanced diet. Use a star rating system to assess the nutritional content of different foods (milk, cereal, chicken or potatoes, for example), awarding between one and five stars to the quantities of protein, fat and carbohydrate each food contains. This is more accessible than comparing mg/100g.

Setting the homework
Ask the children to keep a diary of all the food and drink they consume in one day. It should include everything they eat and drink (including sweets!). Encourage them to ask an adult to complete another copy of the diary so they can compare their results at the end. You may prefer to complete the comparison questions back at school.

Back at school
Read through a sample of the children's diaries (anonymity may be needed in cases of very poor diet). Talk about a balanced diet. Can the children suggest healthier alternatives to cake and sweets? Why do they think more children are overweight than in the past? Anorexia is a topic that could also be touched upon. Some children may be diabetic and, if they are happy to discuss their diet, it may be useful for the class to be aware of their particular needs.

p41 What is digestion? LITERACY LINK

Learning objective
- To know that food is digested by a number of organs, together called the digestive system.

Lesson context
Introduce the structure of the digestive system to the children, and talk briefly about the role of each part of the system (mouth, stomach, small and large intestines) in the process of digestion. It may be worth explaining surface area, and how this can speed up the digestion process. Take a whole potato and compare it with another that you cut into slices. Show the children how the slices cover more of the table when spread flat – this potato would now be digested more quickly because the surface area is greater, and thus easier for digestive enzymes to attack.

Setting the homework
This comprehension activity tests the children's ability to interpret text and draw upon ideas they have already encountered. You may want to read out key words, or even the whole passage together, leaving the children to answer the questions when they get home. You could ask them to write their answers on the back of the sheet, on another sheet of paper, or in their science books/files.

Back at school
Go through the activity with the children. **Answers:**
1. Digestion is about breaking down complicated materials; 2. Food is broken up through slicing and mashing by teeth, chemical attack by enzymes, and breaking up of fat by bile; 3. the small intestine; 4. fibre.

p42 Gills and lungs SCIENCE PRACTICE

Learning objective
- To know the basic structure of the respiratory system.

Lesson context
Look at a picture of the basic structure of the respiratory system with the children, and talk about how it draws in and expels air. Explain that we use only the oxygen from the air drawn into our lungs (we breathe in air containing about 21% oxygen, and breathe out air containing about 16% oxygen). Name the main parts on the diagram – the children need to be familiar with the terms *windpipe* (or *trachea*), *lung* and *diaphragm*.

Setting the homework
Look at the worksheet with the children, and explain that this activity compares the way humans and fish breathe. Tell the children that there are similarities as well as differences, and that this will be a chance for them to find out about these. You may wish to read through the passage with the group, or with children who need extra support.

Back at school
Check through the answers with the class; if time allows, share any interesting facts that the children may have found out. **Answers:** 1. windpipe/trachea; 2. lungs; 3. diaphragm; 4. oxygen; 5. carbon dioxide; 6. its heart.

p44 Does heart rate decrease with age? SCIENCE TO SHARE

Learning objectives
- To know the pulse is produced by the heartbeat.
- To carry out an investigation, make observations and measurements, and draw conclusions.

Lesson context
Talk about the function of the heart, and describe the flow of blood along the major blood vessels (arteries and veins). Ask pairs to check each other's pulse rate over a period of one minute. Explain that the measurements must be taken 'at rest' (even talking or giggling will increase their heart rate). To find the pulse, press the first three fingers firmly along the wrist (or on either side of the windpipe in the neck) until a clear pulse is felt. Children may need a few attempts to gain confidence.

Investigate how activity alters pulse rate by comparing the 'at rest' data with measurements taken after a minute's exercise.

Setting the homework
This activity allows children to look at some of the factors that may affect the heart rate of an individual. Explain that they will be trying to find out whether a person's heart rate slows down with age, then go on to read through the sheet together.

Back at school

Set up a spreadsheet on the classroom computer so the children can type in their results. Use the data to calculate the average results for each age group, then display the results as a graph. Discuss the validity of the children's findings. Results will vary, although young children do tend to have a higher pulse rate than adults. What else might have affected their subjects' heart rate? Talk about state of health, diet (caffeine, for example, stimulates heart rate), state of mind (our heart rate increases if we feel stressed), and any exercise prior to the test. To be definitive, this test would need a far greater number of people, including people of different nationalities, with different lifestyles and so on.

p45 Changes at puberty — SCIENCE PRACTICE

Learning objective
● To know about the changes which take place in the body at puberty.

Lesson context
Discuss the main changes that occur at puberty, within the context of the body becoming sexually mature. Show the children pictures to highlight the differences in body shape brought on by puberty. Emphasise that these changes take place at different rates for different individuals, but that a 'growth spurt', with the development of the sex organs, is perfectly normal and to be expected.

Setting the homework
Your school may well have a newsletter in which parents can be kept up to date with the sex education programme run in PSHE or science. If this is not the case, you may wish to make parents aware of the broad content of this topic so they are suitably prepared for questions. This activity could be given out after a lesson on puberty – explain that it will be a chance for the children to test themselves on what they have learned. They should find this exercise fairly straightforward, although you may want to read it through with less able children, clarifying any new vocabulary.

Back at school

Check through the answers to the worksheet with the class. It would be best if children mark their own work to prevent embarrassment if their answers are incorrect. Talk through any answers that cause debate (for example, in question 12, there could be an argument made for 'both' being the correct answer – while males do become more distinctly muscular at puberty, the female body also becomes physically stronger). **Answers:** 1. B; 2. F; 3. B; 4. F; 5. M; 6. F; 7. B; 8. M; 9. M; 10. M; 11. F; 12. M(B); 13. M.

p46 Smoking — NUMERACY LINK

Learning objective
● To know how tobacco affects the body.

Lesson context
Consider the various uses of drugs – including those used in medicine, and those which are harmful to human health. If you can, introduce the children to a 'smoking machine' (available from educational suppliers) and talk about the dangers and effects of smoking. It is useful for the children to know that while there is often pressure to try smoking, it is an activity that reduces the quality and length of life.

Setting the homework
Some of the children may find this task a little complicated. Talk through the instructions on the sheet, clarifying what they need to do. Make sure they are clear why they should divide the total by 3 (it represents the number of countries in the survey, and so is used when calculating the average).

Back at school
Go through the answers with the class and check the children's bar charts. Talk about some of the difficulties that arise when surveying large groups of people. For example: *Were the people studied all of the same age group? Did they have any other health problems? Had they all been smoking for the same number of years?* Stress that, nonetheless, there is a strong link between the number of cigarettes smoked and the incidence of fatal diseases: if you smoke, the risks increase.

Number of cigarettes smoked each day	Deaths caused by breathing and heart problems: Italy	Deaths caused by breathing and heart problems: UK	Deaths caused by breathing and heart problems: Holland	Total	Average (total÷3)
0	39	52	29	120	40
10	69	81	39	189	63
20	77	98	50	225	75
30	168	177	105	450	150
40	227	248	155	630	210
50	310	325	265	900	300

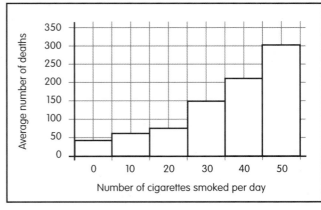

UNIT 2 ANIMALS & PLANTS | LIFE CYCLES

p47 The life cycle of a bean) SCIENCE TO SHARE

Learning objectives
- To know that flowering plants have life cycles.
- To know the factors that affect germination of seeds.
- To practise and develop observational skills.

Lesson context
This activity can take place alongside a series of lessons on life cycles that describe seed dispersal and germination, plant growth and seed setting. It takes between 14 and 18 weeks for a bean to complete its life cycle. You'll need to provide seeds, a pot and, perhaps, compost (use sterile, heat-treated, peat free compost to meet health and safety needs) for each child. It would be useful to grow a set of beans in school in order to demonstrate the difficult parts of the cycle such as transplanting.

Setting the homework
Explain the activity to the children, and tell them that they are going to watch as their beans grow over time. Provide the necessary equipment and, if you are growing a set of beans in the classroom, show the children how to plant the seeds. Suggest ways to record the seeds' growth, for example with a digital camera or by drawing.

Back at school
Ask the children to bring in their worksheet with the pictorial records of the plant's growth at each stage, using the performance of the beans at school as your benchmark. Send the worksheet home again when:
- the bean seeds are planted; the first shoots appear
- when the beans are ready for transplanting (when two strong leaves appear)
- when the first flowers open
- when full pods have formed.

p49 Tree seed detective) SCIENCE PRACTICE

Learning objective
- To know that seeds must be dispersed to help new plants grow.

Lesson context
Tell the children that seeds are the beginning of a plant's life cycle. They must be dispersed in order for the seeds to germinate and grow away from the parent tree. Explain that this can happen in one of three ways:
- heavy seeds (oak and beech, for example) are hoarded by animals as food and may germinate
- seeds encased in berries (rowan and elder) are spread by birds after being eaten
- lightweight seeds (sycamore and birch) are distributed by the wind.

Look at pictures of different trees and seeds, and talk about how these seeds might be dispersed. Explain that seeds are dispersed to ensure that subsequent generations of the plant have access to light, moisture and essential plant foods.

Setting the homework
Make sure that the children understand that there are clues to the method of dispersal in the *description of the seed or fruit* or the *number in each packet* columns.

Back at school
Mark the sheets with the class. The packets containing the largest numbers of seeds are all dispersed by the wind. These seeds are probably smaller and lighter than the others. **Answers:** windblown – maple, birch, sycamore, ash, alder; bird-sown – hawthorn, holly, dogwood, bird cherry; heavy-hoarded – walnut, chestnut, hazel, beech.

p50 Insects and flowers) SCIENCE PRACTICE

Learning objectives
- To know the function of pollen, the stamens and the stigma in fertilisation.
- To know that seeds develop in the ovary.
- To know the role of insects in the process of pollination.

Lesson context
Look at a picture of a flower. Name the key parts of the flower (the stigma, stamens and ovary) and talk about the function of each of these. Talk about the process of cross-pollination and, if possible, go out and collect some pollen from flowers in the local area.

Setting the homework
The children might need to work in pencil, using an eraser to re-order the sentences to correctly answer question 2. Make sure that they understand the meaning of the word *transfer* in question 1c.

Back at school
Go through the correct answers to make sure the children understand the process of cross-pollination. **Answers:** 1a. petal, b. stamen, c. stigma, d. ovary; 2. d, a, e, b, f, g, c; 3. Gardeners might want to encourage insects in order to pollinate flowers and help produce fruits with seeds.

p51 Frogs in the garden) LITERACY LINK

Learning objectives
- To develop an understanding of life cycles.
- To know the life cycle of a frog.

Lesson context
Introduce the children to life cycles, using that of a butterfly to describe the different stages of the life cycle (egg, larva, pupa, adult). A knowledge of the life cycle of different species is used by organic food growers to control pests. Encouraging populations of useful species that feed on crop-damaging pests means that growers need not use expensive pesticides. For example, the ichneumon fly can control cabbage white butterfly populations. The ichneumon fly lays its eggs in the butterfly's larvae as a ready food source for its own hatching larvae, thus killing off the cabbage white butterflies, which can be a pest to cabbage crops.

Setting the homework

Read through both the reader's letter and the editor's reply together.

Back at school

Establish the life cycle of the frog. Explain that at mating the males cling to the backs of the females and fertilise the 2–3000 eggs as they are laid. Only this phenomenal rate of reproduction ensures the species' survival, as many tadpoles and frogs are killed before they mature. **Answers:** 1. Gardeners may want frogs in their garden because they are a natural and effective way of controlling slugs; 2. a, f, d, b, c, e; 3. Enemies of the frog include: (at tadpole stage) fish, newts, water insects and water birds; (at frog stage) grass snakes, rats, hedgehogs, pike, otters and herons.

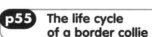

p53 Where has the song thrush gone?   FINDING OUT

Learning objective

● To understand that if living things fail to reproduce they become extinct.

Lesson context

Ask the children to name some 'endangered' species (pandas, for example), and explain that if these species fail to breed, for example because their habitat is disturbed, they may become extinct. Recent human activity in the UK has: increased field sizes on farms by removing hedgerows; improved farm drainage but destroyed invertebrate habitats; increased pesticide use to reduce crop damage. Each of these activities has had a negative effect on the song thrush population.

Setting the homework

Explain that the song thrush is an indigenous (it originates from) British species that was much more common in days gone by. Read through the questions with the class, and stress that they will need to find the answers to the questions from books, magazines or the Internet.

Back at school

Discuss the answers to the questions, and debate the potential benefits to humanity of activities that may threaten other living things. **Answers:** 1. about 22cm; 2. chestnut brown; 3. insects, worms, snails, berries and fruits; 4. They break the shell on a stone or 'anvil'; 5. hedges and thickets; 6. magpies, crows, squirrels, rats; 7. Digging up hedges reduces nesting habitat, and poisoning snails reduces the birds' food supply.

p54 Endangered species in our skies   NUMERACY LINK

Learning objective

● To know that if living things fail to reproduce they become extinct.

Lesson context

Recap on what the children know about 'endangered' species, looking particularly at how if a species fails to breed, for example because its habitat is disturbed, it may become extinct.

Setting the homework

Explain that this homework is about common species of British garden bird. Careful research by ornithologists has shown that the numbers of each species are falling. This data has been adapted from material available on the British Trust for Ornithology's website: www.bto.org

Look at the questions on the sheet, and discuss the meaning of the data in the table. The data in the final column indicates the percentage of sites surveyed at which each species was seen.

Back at school

Ask the children if they can suggest why the number of house sparrows has fallen so sharply. There is no definite answer, but the children might suggest: predation by cats, magpies or rats, or a reduction in food (insects) in urban areas. **Answers:** 1. The children should have crossed out three blackbirds, six house sparrows, five starlings, five song thrushes and two swallows; 2a. house sparrow, b. starling and song thrush; 3a. blackbird, b. house sparrow.

p55 The life cycle of a border collie   SCIENCE PRACTICE

Learning objective

● To know that mammals have life cycles.

Lesson context

Recap what the children know about life cycles, and introduce the life cycle of a mammal, using a cat as an example, ordering the seven stages (which are similar to those listed on the photocopiable worksheet).

Setting the homework

Explain the worksheet to the children. Suggest that they use a dictionary, or ask an adult about the meanings of the words in question 2.

Back at school

Mark the worksheets. Compare answers to question 3, and explain that young animals have a lower chance of surviving through into adulthood than young children in society today. Consequently, other animals need a higher rate of reproduction than humans do for their species to survive. **Answers:** 1. f, d, b, e, a, c, g; 2. weaned – to stop feeding from the mother's milk and begin to feed themselves; embryo – the beginning of life and growing; mature – growing up, ready to take responsibility for the young; 3. Dogs produce so many puppies because, in the wild, few puppies would survive. They may die of hunger, cold or be the prey of another animal.

UNIT 3 THE ENVIRONMENT
WATER AND THE ENVIRONMENT

p56 The water cycle in the kitchen — SCIENCE PRACTICE

Learning objective
● To know that water circulates through the environment in a process called the water cycle.

Lesson context
Look at where water occurs naturally on the planet. Work with the children to establish the processes involved in the water cycle (there is a diagram on page 89 of *100 Science Lessons: Year 5/Primary* 6). Draw and label a diagram of the water cycle, emphasising its cyclical nature – the liquid in the seas evaporates, the gaseous vapour condenses and works its way back to the seas.

Setting the homework
Explain that we can see the water cycle in the home. Point out that the picture on the worksheet shows elements of the water cycle in the kitchen.

Back at school
Make an enlarged copy of the diagram on the worksheet as a visual aid, and discuss the children's answers.

Talk about where the school water supply is obtained from, and where dirty water is treated locally. You may need to explain the function of sewers, sewage works and the U-bend in the kitchen waste pipe. **Answers:** 1. through a system of pipes; 2. the steaming kettle, the boiling saucepan and the drying crockery and cutlery; 3. on the kitchen window; 4. through the extractor fan; 5. through pipes, into the drains; 6. ice in the fridge.

p57 Getting rid of water — FINDING OUT

Learning objective
● To recognise things in our home that remove waste water.

Lesson context
Introduce the idea that people often have too much or too little water around them, and that this can lead to droughts or floods. Hopefully, most children will never experience the catastrophic effects of drought or flood, but our homes are susceptible to the rain and water in the ground. Talk about water conservation and how this can help when water is scarce.

Setting the homework
This activity encourages the children to find out how we remove unwanted water from our homes. Explain that they will need to look carefully at their own homes, talk to their helpers, and look at DIY books to find the answers to the questions on the worksheet.

Back at school
Using an enlarged copy of the worksheet, identify the correct labels for the diagram of the house. You may be able to identify several different kinds of damp-proof course from around the school grounds. **Answers:** 1. C; 2. E; 3. H; 4. A; 5. D; 6. F; 7. B; 8. G; A damp-proof course in a building is close to, but above, ground level, and is a layer of one of the following materials: waterproof rubber or plastic, heavy engineering blue or red bricks, a waterproof wax material injected into holes drilled in the stone or brickwork.

p58 Saving water saves money — LITERACY LINK

Learning objective
● To know that we use a great amount of water and that water must be conserved.

Lesson context
Draw attention to the huge amount of water we use in everyday activities at home and throughout industry (the average family uses 300–700 litres a day). Explain that producing clean, fresh drinking water costs a lot of money and uses up fossil fuels that can damage the environment. Taking water from rivers (and returning waste water to them) damages natural habitats. Ask the children to think how they might be able to conserve water.

Setting the homework
Read through the letter with the children.

Back at school
Discuss the children's answers to the questions. If possible, ask your school's caretaker to explain where the school's water supply comes from, and perhaps give the children a morning and evening meter reading to draw attention to the school's water consumption. **Answers:** 1. Wasting water affects the natural environment and wastes money; 2 and 3. appropriate answers from the list on the worksheet; 4. The readings suggest that 8 litres of water – nearly a bucketful – has been lost through leakage.

p60 A diary of a mini-pond — SCIENCE PRACTICE

Learning objectives
● To create a mini habitat with simple safe equipment.
● To make systematic observation and recording.
● To develop an understanding of the life cycle of some pollution-indicating species.

Lesson context
Talk about how our waste products can pollute natural water sources in the environment. Go pond-dipping, but focus on monitoring a local water resource for pollution indicators such as rat-tailed maggots or water lice. Observe your school's health and safety policy rigorously.

Setting the homework
Explain the sheet to the children, and discuss how they will keep a diary for their mini-pond. Stress that they could use any appropriate container as their mini-pond. Encourage them to think about the dangers of deep water to young children, and the best place to locate their ponds so it will be visited by birds and insects (preferably near plants in a sunny, sheltered location). Suggest that they could develop their mini-pond by adding stones, sand, soil or dead leaves to attract visitors.

Back at school
After five or six weeks, ask the children to bring their diaries back into school to compare. What creatures visited the children's ponds? Which was the most common visitor? Praise those children who have made a systematic record.

p61 Reducing pollution — SCIENCE PRACTICE

Learning objectives
● To recognise some activities that cause air pollution.
● To recognise effects on the environment of air pollution.
● To know some ways in which air pollution can be reduced.

Lesson context

Talk with the children about how burning fossil fuels causes air pollution that may lead to acid rain and global warming. Explain that the latter may cause polar ice caps to melt, raising sea levels and flooding low-lying countries such as Bangladesh. Ask the children if they know of ways to reduce the effects of acid rain and global warming, especially energy conservation, and encourage them to make posters to encourage others.

Setting the homework

Run through the worksheet with the children. Ask them to identify all the electrical appliances in their bedroom as they complete the table.

Back at school

Revise the chain of events that may cause acid rain and global warming. Discuss how we can reduce air pollution by switching off electrical appliances that are not being used. Reinforce this behaviour by developing a similar policy for the classroom. **Answers:** 1. power stations; 2. pollute; 3. acid rain; 4. floods or flooding.

p62 Has the cod had its chips? NUMERACY LINK

Learning objectives

● To understand why an indigenous fish species is in danger of extinction.
● To reinforce that fish stocks need to be conserved if species are to survive.

Lesson context

Ask the children to think about the problems that fish may face because of human fishing and pollution. Draw attention to the fact that pollution, and the overfishing of species lower down the food chain, are affecting fish stocks. Some fish species are in danger of extinction and some species need to be conserved to survive.

Setting the homework

Make sure the children can read the large numbers in the table. Draw their attention to the diagram that shows them how to draw the graph.

Back at school

Recap the answers to the questions with the children. Explain at the end of the session that fishermen are not allowed to catch so much cod now because the species is in danger of extinction. **Answers:** 1 and 2. The graphs show a decline in tonnage caught until 1991, a small increase in 1993 and 1995 and then further decline; 3. There are several reasons, including: overfishing (too many fish are being caught), pollution of North Sea waters, global warming (the surface temperature of the sea *may* be increasing and this *may* affect the developing fish in their infant state).

These figures have been adapted and rounded to the nearest 20 000 tonnes from information available on the Centre for Environment Fisheries and Aquaculture website (www.cefas.co.uk).

UNIT 4 MATERIALS | GASES, SOLIDS AND LIQUIDS

p63 Gas to liquid SCIENCE PRACTICE

Learning objectives

● To reinforce that some materials can be classified as solids, and some as liquids.
● To know that air is a type of a material called a gas.
● To know that air has mass.

Lesson context

Review work from Year 4/Primary 5 on solids and liquids and their properties. Consider *air* as a material. Show the children a *balloon balance* (two similar balloons tied to a coat-hanger as if on a set of scales) and ask them to predict what will happen when you let air out of one balloon. Let the children see whether their prediction was correct by (slowly!) deflating one balloon. Explain that sometimes balloons can be filled with helium, which is also a gas – balloons filled with helium are the ones that float off if you don't hold on to them! Ask: *Is helium lighter or heavier than air?* (Lighter.)

Setting the homework

Read through the sheet, making sure that the children understand what they have to do. Recap the balloon experiment if necessary for question 2.

Back at school

Review the children's responses. Talk with the children to make sure they understand that air is a gas, and can explain how gases differ from solids and liquids. **Answers:** 1. solid, squashed, shape, liquid, runny, squash, shape, gases; 2a. see diagram below:

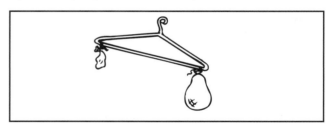

b. Balloon 'Y' contains air and this is what makes the balloon heavy/heavier; c. Use the same type and size of balloon, and the same number of pumps of air in each; balloon; d and e. see diagrams below:

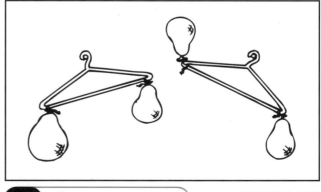

p64 Air FINDING OUT

Learning objectives

● To know that air is all around us.
● To know that air is a type of material called a gas.

Lesson context

Demonstrate some of the properties of air. Pump up a beach ball to show that the air inside it expands to fit the space available and exerts a pressure on the walls of the ball. Show that air has mass by making a simple balloon balance (tie two balloons to a coat-hanger 'balance' to show that balloons can weigh more than one another). Discuss and demonstrate 'air resistance' (for example, drop a toy man from a height of 2m with and without a parachute – which falls faster? Air resistance slows down the toy man with the parachute.) Then go on to identify and discuss devices that make use of the air (kites, car tyres, windmills, for example).

Setting the homework

Explain the task to the children, and how long you expect their reports to be (this can be decided according to the children's ability – maybe up to a page for more able children). Talk about where they might be able to find the necessary information (from the local or school library, or the Internet). Emphasise the importance of note-taking – not copying text – and the features of a non-chronological report, such as introducing the subject and describing in the present tense. Encourage pictures or diagrams. The children should make notes before writing their report.

Back at school

Share the children's reports. Ask some children to talk about any interesting facts they have included in their report. Use the reports for display work or a class book.

p65 Solid, liquid or gas?   SCIENCE TO SHARE

Learning objectives

● To know that gases do not have a fixed volume as solids and liquids do.
● To know that gases are more easily compressed (squashed) than solids and liquids.
● To know that gases and liquids can flow and change their shape, but continuous solids have a fixed shape.

Lesson context

Explain the different properties of solids, liquids and gases to the children. Show how a solid (a brick, for example) cannot change shape or volume; a liquid can change shape but not volume, and a gas (using an inflated balloon), can expand, pour and flow.

Encourage the children, through role-play, to think about the arrangement and movement of particles in a solid, a liquid and a gas.

Setting the homework

Tell the children that adult supervision is necessary for this activity. Point out the warning on the photocopiable sheet, and warn that they should never take lids off bottles if they don't know what's in them, especially if they are kept in a locked cupboard, or the garage or garden shed.

Back at school

Display an enlarged copy of the chart on the board. Let the children share their findings and thoughts with the rest of the class, filling in the chart with their answers. Through discussion, establish that: solids have a fixed shape, cannot be squashed, do not spread out into the air when released and are not runny; liquids do not have a fixed shape, cannot be squashed, do not spread out into the air when released, but are runny; gases do not have a fixed shape, can be squashed, do spread out into the air when released and are runny.

The completed diagrams should look like this:

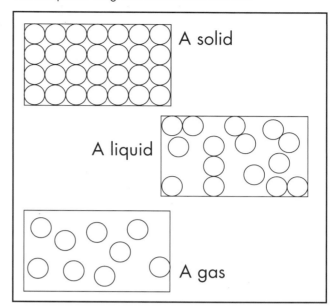

p66 Steamy   SCIENCE PRACTICE

Learning objectives

● To know that gases can be turned into liquids by cooling.
● To know that this process is called condensation.
● To know that water vapour is present in the air, but cannot be seen.

Lesson context

Recap on previous work dealing with the three states of matter (solid, liquid and gas), and particularly the process of evaporation. Discuss and explain what happens during condensation, referring to everyday examples (such as droplets on the side of a cold glass, or mist on the bathroom window). Demonstrate by refrigerating a glass and allowing it to warm during the lesson. Reinforce melting/freezing and evaporation/condensation, and how these processes are reversible.

Setting the homework

Read through the worksheet with the children. Make sure they know how to answer questions 5 and 6.

Back at school

Mark the worksheets together and recap the process of condensation. **Answers:** 1. heat it up; 2a. liquid, b. gas; 3. cool it down; 4. condensation; 5. C; 6. A and C are examples of condensation.

p67 Liquid to gas   NUMERACY LINK

Learning objectives

● To know that when a liquid is heated to a certain temperature, it boils.
● To know that when a liquid boils, it changes into a gas.
● To know that water boils at 100°C.

Lesson context

Ask the children what they think happens to the temperature of water as it is brought to the boil. Explain that you are going to do an investigation to find out what happens. Demonstrate the experiment to the children, boiling a kettle of water and using a data-logger to record the rising temperature as the water boils.

⚠ It is essential to keep any boiling water well away from the children to minimise the risk of scalding. If you have no data-logging equipment available, simply use the data on the sheet to set the scene for the children.

Setting the homework
Discuss the table of results, and ask the children what would be the best way to present this information (they need to draw a line graph, not a bar chart). Discuss how each axis should be labelled.

Back at school
Look at the children's completed graphs. You might find it worthwhile to recap how to draw a graph using the data on the homework sheet.
Answers:

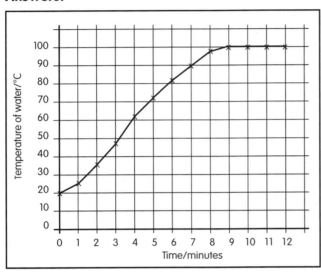

p68 What a state!   LITERACY LINK

Learning objective
● To know that heating/cooling can cause materials to melt, boil, evaporate, freeze or condense.

Lesson context
Demonstrate how to make wax candles using a candle-making kit. Explain how heat is used to melt the wax so it can be shaped, and when it cools it 'freezes', and becomes solid again. Ask the children to think about how heating can speed up evaporation (such as puddles on a warm day) and how rain is the result of water vapour condensing in clouds.
⚠ Take special care when working with naked flames in the classroom – make sure that long hair is tied back and loose clothes are tucked away. It may be better to carry out this activity as a demonstration, with the children sitting away from the flames. (See page 102 of *100 Science Lessons: Year 6/Primary 7* for more information on health and safety in the classroom.)

Setting the homework
Read through the worksheet together, and explain that as well as testing their knowledge you are also looking for accurate spelling and the correct use of these important science words.

Back at school
Mark the worksheets together, and talk about any answers the children are unsure of. **Answers:** solid, liquid, gas, Solids, liquids, gases, Solids, liquids, gases, Gases, liquids, solids, melting, boiling.

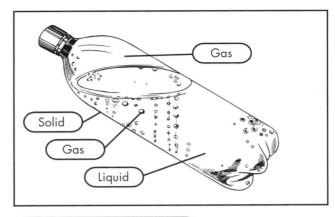

p69 Solid to liquid   SCIENCE PRACTICE

Learning objectives
● To conduct an investigation into rates of melting.
● To know that the freezing/melting temperature of water is 0°C.
● To know that the temperature in the classroom is normally 18–22°C.

Lesson context
With the children, carry out an investigation to find out how long ice keeps a drink cold. Using a data-logger (or a thermometer – see page 90 of *100 Science Lessons: Year 6/Primary 7* for information about using thermometers safely), measure the temperature of a glass of ice-cold water at regular intervals until the ice has melted and the water has reached room temperature. Consider what factors might affect the rate of melting (such as the temperature of the room, or the insulating properties of the container the drink is in). The water will stay at around 0°C until all the ice has melted.

Setting the homework
Look at the table of melting point values on the worksheet. Discuss how the children might decide on a scale for their bar chart, and what you expect from a bar chart such as labelled axes, a title, and for the chart to be accurate, tidy and clear.

Back at school
Go through the answers to the sheet. Look at the children's completed bar charts.
Answers: liquid, melts, melting point, freezing point; 1a. ice, b. iron; 2. ice; 3. aluminium; 4. It will turn into a gas; A completed graph:

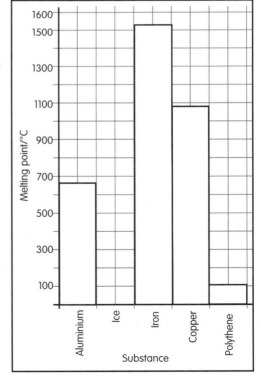

UNIT 5 ELECTRICITY | MAKING AND USING ELECTRICITY

TEACHERS' NOTES YEAR 5

p70 Going on and on and on... **NUMERACY LINK**

Learning objective
● To know that in a battery, different chemicals are brought together to produce electricity.

Lesson context
Discuss where we get electricity from, and how we can tell when electricity is present (perhaps by seeing a light or hearing a sound). Talk about batteries and where we might find these in everyday life (for example, toys and wristwatches).

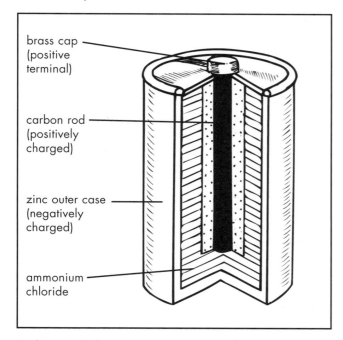

brass cap (positive terminal)

carbon rod (positively charged)

zinc outer case (negatively charged)

ammonium chloride

Explain simply how batteries work. Spend some time discussing the 'life' of a battery, and that a 'dead' battery occurs when the chemicals in the battery have been used up.

Setting the homework
This homework looks at the 'life' of a torch battery (look out for battery adverts that compare batteries like this). The children may need reminding how to draw a bar chart. The grid provided allows for 1:1 scales on the axes if the bars are drawn hoizontally or, better, a 1:2 scale if the bars are drawn vertically. Encourage more able children to use a 1:2 scale.

Back at school
Check that the bar charts have individual columns and that the axes and graph itself are labelled correctly. **Answers:** 1. Forevercell; 2. four; 3. the Everuse pack.

p71 Locked out **SCIENCE PRACTICE**

Learning objective
● To know how an electromagnet can be constructed and used.

Lesson context
Construct electromagnets by winding many turns of insulated copper wire around a large iron nail. Attach each end of the wire to a terminal of a battery.

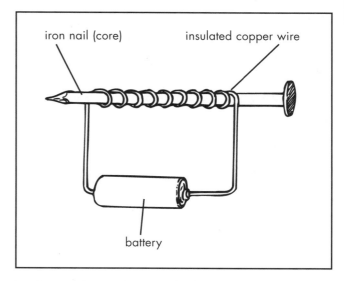

iron nail (core)   insulated copper wire

battery

Test how to change the strength of an electromagnet by changing the number of coils of wire or the strength of the battery. Ask the children to use secondary sources to look at some common uses of electromagnets.

Setting the homework
Make sure that the children know how an electromagnet works. Remind them which materials are magnetic and which are not.

Back at school
Discuss the answers with the children. **Answers:** 1. a circuit; 2. electricity/an electric current; 3. It is attracted to the electromagnet; 4. Aluminium is not a magnetic material; 5. The lock would not work; 6. Add a bulb or buzzer next to the switch.

p72 Spinning around **SCIENCE TO SHARE**

Learning objective
● To know some uses of electric motors.

Lesson context
Show the children how an electric motor operates (those from toy racing cars are ideal) and how it can be made stronger or more powerful. A larger motor is usually more powerful because it has a thicker wire (which can carry more current), more coils and stronger magnets to create a stronger magnetic field.

Setting the homework
Explain the sheet to the children. Remind them that they should not take apart electrical items as they could injure themselves and/or damage the appliance.

Back at school
Compare the children's lists, and consider which are the largest or most powerful motors (for example, washing machine motors) and which are the smallest or least powerful (perhaps those in personal stereos or toy cars).

p73 Simon's new crane **SCIENCE PRACTICE**

Learning objectives
● To use electric motors in toys.
● To understand circuit drawings including motors.

Lesson context
Using small motors, batteries, wires and collage materials, allow the children to design and make simple toys that contain a motor, such as a robot with a spinning head, a windmill or a toy vehicle. Introduce the children to the idea of drawing a circuit diagram to show what they have made. The circuit diagram symbol for a motor is: ─(M)─

Setting the homework
Explain the worksheet to the children. This activity requires the children to describe the action of a circuit from a diagram of it. Make sure they know what all the symbols represent.

Back at school
Discuss how the children think the crane might work. You might like to build a model crane using the circuit diagram to demonstrate how each switch moves it.

p74 We have the power **FINDING OUT**

Learning objective
● To realise that power generation has environmental implications.

Lesson context
Explore how electricity is generated. 'Discover' that different types of power station exist, such as coal-fired or hydroelectric, and look at the different ways in which electricity is produced in each of these power stations.

Setting the homework
Give guidance on where the children might look for information (such as the library or the Internet). Encourage them to make notes from their research – such information management is a key literacy skill at Year 5/Primary 6.

Back at school
Give the children the opportunity to 'show and tell' what they have found out, then allocate groups to collate and represent their findings as material for display.

p75 Emission control **LITERACY LINK**

Learning objective
● To know that 'environmentally friendly' power sources exist.

Lesson context
Let the children research renewable energy sources to promote a discussion about environmentally friendly energy supply and how we can help to conserve energy in school and at home.

Setting the homework
By the end of Year 5/Primary 6, children should be familiar with persuasive writing activities. This cloze text ties together the children's work on mains power supply and prepares them for any upcoming unit assessment by reinforcing key vocabulary.

Back at school
Check the children's answers. Each word is only needed once. **Answers:** burning, generator, fuel, carbon dioxide, Sun, air, rivers, waves, tides, pollute.

UNIT 6 FORCES & MOTION | EXPLORING FORCES AND THEIR EFFECTS

p76 Mind the gap **FINDING OUT**

Learning objective
● To know that a variety of structures are used to support the weight of objects.

Lesson context
Look at pictures of different bridges and note differences in their construction. Using rolled-up paper or construction kits, ask the children to compare the strength of different-shaped beams (round, square or L-shaped), and to investigate how these can be formed into arches. Encourage the children to look out for strong and weak points in the construction of their structures.

Setting the homework
This research activity asks the children to look at the real-world uses of structures and how arches and beams work together to complete large structures, in this case bridges. Talk about possible sources of information with the children. Encourage them to summarise the information they find and bring it back to school.

Back at school
Ask the children to share any information they found. This homework could be used to provide material for use later in projects or posters for display alongside the children's bridge models and investigation work.

p77 A day out at Adventure Island **LITERACY LINK**

Learning objective
● To revise that friction can be reduced by lubricants.

Lesson context
Pull a wooden block attached to a piece of string across the desk. Ask the children to describe the forces of friction acting as you pull, then to think about how they could reduce the effect of friction. The children should suggest that the smoothness of a surface affects the force of friction; they should also know how lubricants can reduce the force of friction. Let them investigate which substances make the best lubricants (water, oil and washing-up liquid are good things to try).

Setting the homework
This homework takes the idea of how surfaces affect friction into a natural environment. Talk about the children's experiences of slippery and rough surfaces (such as ice and tarmac) as you read through the worksheet.

Back at school
Go through the children's responses to the activity. Do they have any experience of the surfaces described on the sheet? How were they affected by them – were they helped or hindered? **Answers:** A.  Crusoe's Beach: high friction – sand, dry rocks, shells; low friction – seaweed, water on rocks; B. Adventure Forest: high friction – tree bark; low friction – smooth, waxy leaves; C. Tree town: high friction – ropes; low friction – wooden steps; D. Deep River: high friction – gravel beach; low friction – slimy fish, wet path, algae or weed on stones, wooden bridge (especially if wet).

p78 Loosen up — ASK AN ADULT

Learning objective
● To know that friction can be reduced by lubricants.

Lesson context
Remind the children that lubricating a surface can be of benefit (such as using valve oil on a brass musical instrument), but also that too much lubrication – too little friction – can lead to danger (such as water on a road surface causing a skid).

Setting the homework
Ask the children to act as reporters, and to collect stories of how friction and (liquid) lubricants have been used by friends or family.

Back at school
Invite the children to share their stories. The readings could be recorded and put together on video to be used as an assessment tool, or for use with younger children.

p79 Ball-bearing clocks — SCIENCE PRACTICE

Learning objectives
● To know that everything falls down due to gravity.
● To know that falling materials can be used to measure time.
● To identify features of a gravity-controlled timer.

Lesson context
Show the children examples of sand and water clocks. Explain how gravity affects the falling material, and how the rate at which the medium falls can be altered to change the time taken for the clock to run out (for example, by altering the size of the hole the sand trickles through). Let the children make their own timers using collage materials.

Setting the homework
Explain the sheet to the children. They will need to understand that gravity affects all objects, and have some experience of simple levers (such as the see-saw).

Back at school
Check the children's answers. **Answers:** 1. gravity; 2. friction; 3. It pushes up a lever/see-saw; 4. three; 5. Five minutes.

p80 Sail racing — NUMERACY LINK

Learning objective
● To know that the force of the wind can be used to provide energy for work.

Lesson context
Talk about windmills, and explain how wind provides a force, which can power objects that have sails to catch the wind. Ask the children to think about windmills grinding corn, and emphasise their similarities to wind turbines.

Setting the homework
The homework extends the idea of wind as a source of energy to look at boat sails. Remind the children how to calculate areas given the width and height of a rectangle, and of the features of a line graph. For less able children you may wish to calculate the areas together in advance, complete the table before copying the sheets, or make the sails with centimetre squared paper and let the group use counting methods rather than calculation to find the areas.

Back at school
Check the children's answers – look for correctly marked linear scales and labels on the axes. **Answers:** Area of sail (cm²) 25, 50, 100, 200, 400. The completed graph should look like this:

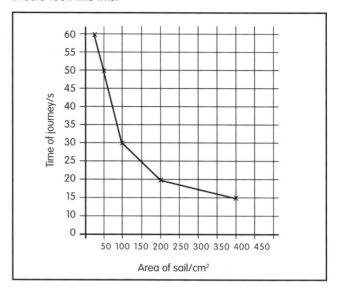

p81 A sudden jump — SCIENCE PRACTICE

Learning objective
● To know that energy stored in an elastic band can be turned into movement energy.

Lesson context
Make elastic band 'crawlers' using cotton reels, elastic bands and matchsticks. Ask the children to investigate how the tension in the elastic band is converted into movement energy – they should discover that different thicknesses of elastic band

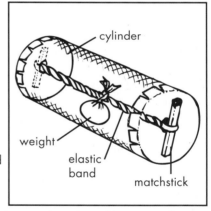

develop different tensions and produce different results – encourage the children to investigate what happens with different 'motors' in their crawlers.

Setting the homework
The effect of converting tension (elastic potential energy) into movement is equally as applicable to springs as it is to elastic bands – try taking a retractable ballpoint pen apart to show the children where energy can be stored. This activity investigates this as a practice/revision exercise.

Back at school
Check through the children's answers, and discuss and correct any misconceptions. **Answers:** 1. Because the suction of the sucker at the bottom holds the toy down until the upward force of the spring breaks the air seal; 2. the spring inside it; 3. gravity; 4a. The force from the spring would be stronger and the toy would jump further, b. The force from the spring would be less, and the toy would not jump as far; 5. There must not be too much weight for the force in the spring to push up into the air.

UNIT 7 LIGHT & SOUND | BENDING LIGHT AND CHANGING SOUND

p82 Crooked pencils | SCIENCE TO SHARE

Learning objectives
● To know that when light rays pass through a curved, transparent material the paths of the rays are changed.
● To make observations and draw conclusions.

Lesson context
Show the children examples of transparent, translucent and opaque materials, and explain the differences between the three types of material. Ask the children to record, in pictures and writing, the appearance of objects observed through transparent materials that differ in thickness across their width (such as a magnifying mirror that appears to enlarge objects).

Setting the homework
Explain the sheet to the children. As well as light being affected by the thickness of the material it is passing through, when light travels from one medium into another its speed and direction changes. This is known as refraction, and leads to the apparent difference in depth of materials as well as the lens effect the children have seen in class. Emphasise safe practice if the children choose to extend the homework with observations at the swimming pool.

Back at school
Look at the children's work. They should have drawn a diagram showing the straight object bending or enlarging (or both) in the glass as it goes from the air into the water. The coin under the glass should appear larger or nearer.

p83 Who wears the glasses? | NUMERACY LINK

Learning objective
● To know that the eye contains a lens and a place where images form.

Lesson context
Extend the children's work on lenses to looking at the structure of the eye, identifying the important parts (the pupil, iris, retina and optic nerve) and their functions, particularly the lens. Explain how long- and short-sightedness are corrected by using a lens to focus light correctly on the back of the eye. Research, with sensitive reference to class members, defects of vision and how these are corrected using lenses.

Setting the homework
Link this activity to data-handling in maths. More advanced surveys could require the children to survey family and friends who wear glasses to find out if they are long- or short-sighted.

Back at school
Look at the data collected – does it show a general trend across the class? Combining the children's surveys will give a more statistically valid sample. The data could be entered into a computer spreadsheet package to do the adding up and the plotting of more complex graphs – particularly of pie charts, which are likely to be beyond most Year 5/Primary 6 children.

p84 Primary colours | FINDING OUT

Learning objective
● To know that sunlight is made from a range of different-coloured light rays.

Lesson context
Show the children how light can be split into its seven component colours using a prism. Ask the children to make multi-coloured spinners to show how white can be seen when a spinner with all the colours of the spectrum is spun at high speed.

Setting the homework
Explain that this is a 'finding out' activity, and direct those children who may not have access at home to reference books or to the local library to appropriate resources in school.

Back at school
The different definitions of 'primary' colours can cause considerable confusion for children. If possible, demonstrate both types of colour mixing to confirm the children's answers. **Answers:** artist's primary colours – red, yellow, blue; artist's secondary colours – orange (red + yellow), green (yellow + blue), purple (red + blue); scientist's primary colours – red, green, blue; scientist's secondary colours – yellow (red + green); magenta (red + blue); cyan (green+blue).

p85 Ding dong | SCIENCE PRACTICE

Learning objective
● To know that vibrating objects produce different sounds.

Lesson context
Let the children try making different sounds using simple everyday objects ('twanging' rulers, elastic bands or upturned containers) and some musical instruments. Establish that the part of each object that is vibrating has made the sound.

Setting the homework
Talk briefly about some sources of sound in school and the environment. Ask the children to listen silently for one minute and list everything they can hear. Explain the worksheet to the children.

Back at school
Discuss the children's answers, particularly with respect to PSHE for question 5. Focus on sounds in the environment that give messages, such as the bell at break time, phones and doorbells at home, the bleep on a Pelican crossing, or the siren on an emergency vehicle. Talk about the health and safety issues highlighted by the activity – how should we act if we hear the fire alarm? (Children should listen for and follow the teacher's instructions immediately.)
Answers: 1. alarm bell, voices, door slamming, footsteps; 2. the alarm bell and the teacher's instructions (you can debate this one!); 3. Electricity makes the clapper strike the bell, and the bell vibrates; 4. Sound vibrates air inside the ear, which vibrates parts inside the ear (more able children may find out the names of the bones here) and is sensed by a nerve that carries a message to the brain, which translates into the sound we hear.

p86 I can't hear you! — NUMERACY LINK

Learning objectives
● To know that sound travels in straight lines.
● To know that sound can travel through air.

Lesson context
Explore how sound travels through the air. Ask the children to predict, and then find out, what will happen when they change the distance of a detector from a sound (to recap previous work and support the homework). Investigate how sound intensity varies with the angle of a detector to the sound. Use the children's results to plot a graph of what they found out. What are the implications of this for animals whose ears can swivel round? (They are more able to identify the direction of a sound source – be it predator or prey!) Are the children aware of turning to listen?

Setting the homework
Explain to the children that the correct unit for sound intensity – that they are probably calling volume – is the decibel (**dB**). Explain that they are going to look at a graph similar to the one their investigations will have produced, but this time the distance is changing rather than the angle.

Back at school
Check the children's graphs, looking particularly at use of scale and labelling of axes. **Answers:** Yes, her mum was right: the sound level is higher nearer to the TV. To make this a fair test, the sound made by the TV should be the same (for example, the same pop song), the volume must not be altered, and the sound meter should always point in the same direction.

p87 My music — LITERACY LINK

Learning objectives
● To know that the pitch of some musical instruments can be altered by changing the size or tension of the vibrating part.
● To know that the loudness of musical instruments can be altered by changing how much they vibrate.

Lesson context
Examine how the tension, length or thickness of an elastic band affects the pitch of the note it makes when 'twanged'. In groups, let the children design and make musical instruments. Can they tell you what changes the pitch or loudness of the notes in each instrument? Can they play a *recognisable* tune together?

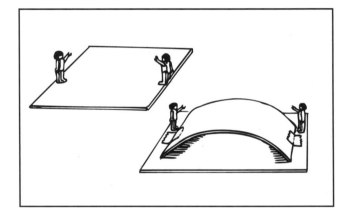

Setting the homework
This activity is about using vocabulary creatively to describe sounds, so it might be worthwhile discussing describing words with the children so they have a greater range of words to choose from. Encourage at least a few children to bring in the tape or CD that has inspired their writing.

Back at school
Ask some children to read out their descriptions as you play the pieces of music to the class. Do the rest of the class agree with the comments?

100 SCIENCE HOMEWORK ACTIVITIES ● YEAR 5

UNIT 8 EARTH & BEYOND | SUN, MOON AND EARTH

p88 Is the Earth round? — LITERACY LINK

Learning objective
● To know that the Sun, Earth and Moon are approximately spherical.

Lesson context
Demonstrate 'flat Earth' and 'round Earth' models to the children, using play figures to show how people can 'disappear' over the horizon, suggesting that the Earth is round. Discuss the anecdotal evidence that the Ancient Greeks observed to suggest that the Earth is round: star movements, the way ships disappear over the horizon, and the circular shadow cast by the Earth in a lunar eclipse. Is our evidence today for a round Earth direct or indirect? (Direct – pictures from space since the 1950s and 1960s.)

Setting the homework
You may wish to carry out this lesson alongside a history lesson on 'Explorers', looking at the life of Christopher Columbus.

When discussing the homework, set the scene and try to get the children 'into character' before they go home. Explain what you want the children to produce – this could either be a simple writing exercise, or it could be extended to an art and design project if you would like the children to produce a creative document.

Back at school
Allow the children to read out some of their writing. This is an excellent opportunity for role-play, especially if costumes are available. The completed court circulars will make a good display.

p89 Day and night · SCIENCE PRACTICE

Learning objectives
● To use scientific knowledge to explain observations.
● To know that the Sun's apparent movement across the sky is caused by the Earth spinning on its own axis.

Lesson context
Recap what causes day and night, then ask the children to make a human model of how this happens. Arrange them like this:

children representing the Earth

child representing the Sun

The children represent different places on Earth. If a child can see the Sun (without turning their head) it is day.

This can be repeated with a torch and a globe. Talk about how and why the number of daylight hours changes throughout the year.

Setting the homework
Read through the worksheet with the children, making sure that they understand the diagrams.

Back at school
Ask individual children for their answers. For questions 3 and 4, invite possible answers until no other suggestions are forthcoming. **Answers:** 1. 24 hours; 2. 365¼ days; 3a. midday, b. midnight, c. sunrise; 4. Look for an arc made by the Suns, with the highest point at midday; 5. Because the Earth spins on its axis.

p90 Daylight and darkness · OBSERVATION

Learning objectives
● To use scientific knowledge to explain observations.
● To know that the Sun's apparent movement across the sky is caused by the Earth spinning on its own axis.

Lesson context
Once children have a firm grasp of what causes night and day, ask them to design a crossword or wordsearch using key vocabulary. Less able children could adapt a ready-made crossword (for example, from *100 Science Lessons: Year 5/Primary 6*, page 194). Write in the first one or two letters of each answer.

Setting the homework
It might be useful to set this homework after the October half-term break, once the clocks have gone back and darkness begins to set in earlier. This activity will take four weeks to complete, with children recording the sunset time twice a week during this period. Explain what they should do and when you expect the homework to be handed in (they might need reminders throughout the month!). Suggest that recording sunset times on a dull day might be difficult; if there has been a spell of dull weather, remind the children to make their observations on a day when the Sun does come out. Discuss how, in order to obtain reliable recordings, children will need to use the same measure to determine when sunset has occurred (for example, noting the time a street light comes on).

Back at school
Use one child's set of results as a model. Write their results in a prepared table, either on the board or OHP, then draw a line graph of these results (you could have the axes already drawn on an OHT). The rest of the class should then plot a line graph of their own results. Less able children might find it easier to draw a bar chart of their results.

p91 Making a sundial · SCIENCE TO SHARE

Learning objectives
● To know that time on a clock varies around the world.
● To see how some units of time are linked to the motion of the Earth, whereas others are artificial.

Lesson context
Discuss the meaning of some time words. Divide words into those that are 'natural' (related to the Earth's movement), and 'artificial' (arbitrary human divisions such as the hour or minute). Look at how time zones are related to the rotation of the Earth, and encourage the children to think about how time zones affect travellers. Have any children experienced jet lag? Can they explain it to the class?

Setting the homework
This is a weekend or holiday task, and would best be set in the summer term, when there is (hopefully) more chance of there being a sunny day (the Sun will be higher in the sky too, which means more people's gardens will have more sunshine). Be sensitive to the fact that some children may not have a suitable outdoor space to allow them to carry out this activity – you could help children set up their experiment in the school grounds.

It may be useful to demonstrate this activity in class, which will also provide some back-up data for children who can't take readings. Emphasise the health and safety issues: the need to take care so that others do not hurt themselves on the sticks and string in the ground.

Back at school
Ask some children to tell the rest of the class what they found out. Some may wish to draw their shadow pattern on the board. The results of this activity, and the children's writing about what they did, would make a good display. This is the kind of pattern to expect:

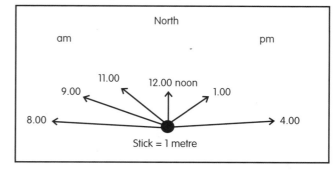

North
am · pm
11.00
9.00 · 12.00 noon · 1.00
8.00 · 4.00
Stick = 1 metre

p92 Day-lengths — NUMERACY LINK

Learning objectives
● To know that other planets have different day-lengths from Earth.
● To think logically about causes and effects.

Lesson context
Discuss with the children what they understand by the term *day*. Clarify that one day is the time it takes a planet to rotate once on its axis. Introduce the children to data showing how long a day is on the other planets in our Solar System (see, for example, the table on *100 Science Lessons: Year 5/Primary 6*, page 186). Ask them to imagine what it would be like living on a planet with a different day-length.

Setting the homework
Ensure that all the children know how to work out the time, in Earth days, that it takes a planet to make one spin on its axis. For example, a Mercury day is 1400 Earth hours; 1400 ÷ 24 = 58 Earth days. Emphasise that they will need to put their answers for question 2 into the table of data.

You will need to provide graph paper for question 3. Help less able children label their axes and choose a scale to use on the vertical axis of the graph before taking the sheet home. Emphasise the need for accuracy and neatness when drawing graphs.

Back at school
Check the children's answers. **Answers:** 1. 24 hours; 2. Mercury – 58 days, Venus – 241 days, Earth – 1 day, Mars – 1 day 1 hour, Pluto – 6 days; 3. bar chart:

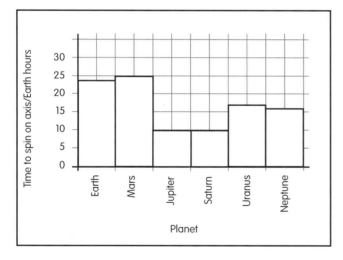

p93 The four seasons — SCIENCE PRACTICE

Learning objectives
● To know that the Earth travels in an orbit around the Sun once a year, and this causes the seasons we experience.
● To know that the pattern of seasons is related to the Earth's tilt and its orbit around the Sun.

Lesson context
Model the Earth's orbit around the Sun using a globe and torch. Point out how at certain times the UK is tilted towards the Sun and at other times is tilted away, and that this is what causes the seasons. Ask the children to tell you when it is hot in the northern hemisphere (summer) and when it is cold (winter), and why this is. (In summer, sunlight falls on a much smaller area because of the tilt of the Earth, and so its heat is more concentrated.)

It might also be useful to show the position of the North Pole at different times of the year, so that the children will have a better chance of answering question 4 correctly.

Setting the homework
Make sure the children understand what they have to do. You may need to explain the diagram for question 3, so that the children appreciate when the UK is tilting towards and away from the Sun.

Back at school
Ask children to share their answers. **Answers:** 1a. spring, b. summer, c. autumn, d. winter; 2. We have different seasons because the Earth's axis is tilted. The axis is tilted at 23.5°; 3a. summer, b. autumn, c. winter, d. spring; 4a. summer, b winter.

p94 The Earth's 23½° tilt — LITERACY LINK

Learning objectives
● To know that the Earth travels in an orbit around the Sun once a year, and this causes the seasons we experience.
● To know that the pattern of seasons is related to the Earth's tilt and its orbit around the Sun.

Lesson context
Remind the children how the tilt and movement of the Earth causes the seasons. Discuss what would happen if the Earth was tilted so that the North Pole was pointing directly at the Sun – what effect would this have on temperature in different parts of the world? Ask the children to use secondary sources to find out about animals that could live in permanent light and heat (reptiles or camels, for example), and those that could live in permanent dark and cold (bats or sea life).

Setting the homework
Tell the children that when they answer question 2 they must use their imagination as well as their scientific knowledge to write a report. If they wish, they can draw pictures too.

Back at school
Ask some children to read out their account of what life would be like if the Earth was not tilted. **Answers:** 1. Earth, tilted, towards, higher, longer, away, lower, fewer; 2. No tilt means: no seasons; every day of the year will have the same number of hours of daylight and darkness; the daily temperature throughout the year in the UK would vary little compared to what we experience with the changes of the season.

UNIT 1 OURSELVES NEW BEGINNINGS

p95 Family fitness plan SCIENCE TO SHARE

Learning objective
● To construct a plan for a healthy lifestyle.

Lesson context

Examine aspects of a healthy lifestyle. Ask the children to think about a typical day, and to break it up into the various activities that form part of their own lifestyle. Ask them to consider which activities promote good health, and which aspects are not so healthy.

Setting the homework
Explain that this homework will provide the opportunity to involve family members in planning activities that will improve the whole family's health and fitness by designing an activity plan. Tell them that they will need to plan activities that can become part of the family's daily routine. A simple idea might be to go without the TV remote control for a week – the frequent movement, bending and stretching will help to limber up the body! Encourage the whole family to keep a diary, putting a star on every day an activity has been completed.

Back at school
Share the children's ideas. As a group, consider which ideas seem realistic – try to praise and encourage even minor changes in lifestyle. Think about the health benefits each activity might bring.

p96 Lifestyle muddle SCIENCE PRACTICE

Learning objective
● To identify activities and substances that may be harmful to health.

Lesson context
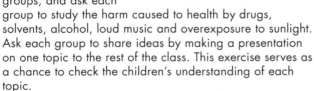
Divide the class into small groups, and ask each group to study the harm caused to health by drugs, solvents, alcohol, loud music and overexposure to sunlight. Ask each group to share ideas by making a presentation on one topic to the rest of the class. This exercise serves as a chance to check the children's understanding of each topic.

Setting the homework
This activity follows on directly from the work covered in class. Explain that the children need to match each activity with its effect on the body.

Back at school
Mark the worksheets as a class, and check that the children are happy with the answers. Clarify any areas of uncertainty that may have arisen. **Answers:** Eating too much sugar and fat – can lead to obesity (being seriously overweight) and put stress on the heart...; Taking exercise – strengthens the heart and other muscles...; Fibre – found in vegetables, fruit and cereals...; Recreational drugs – these are not prescribed by a doctor...; Sunbathing – UV light causes skin cancer...; Cigarettes – can cause heart and respiratory diseases, including cancer; Loud music – can cause hearing loss; Medicinal drugs – are prescribed by a doctor, and help to fight illness; Solvents – inhaling their vapours can cause brain and liver damage.

p97 Staying alive! NUMERACY LINK

Learning objectives
● To know the stages of the human life cycle.
● To assess the healthy and unhealthy aspects of a life style.

Lesson context
Recap on the children's previous work on healthy lifestyles, and on the stages of the human life cycle from fertilisation, through birth, infancy, childhood, adolescence to adulthood. This activity is designed to test numeracy skills, as well as revealing the harsh reality of infant mortality rates, which are still high in many poorer countries.

Setting the homework
Make sure the children are confident at working out percentages. If possible, work through an example with them and perhaps ask them to make one of the calculations on the back of the sheet to reinforce the equations. Remind them how pie charts can be used to display percentages.

Back at school
Work through the answers with the children, and check they understand where errors have occurred. It might be useful to discuss the results with them – why do they think fatal childhood diseases are more common in poorer countries? (Less reliable sanitation and clean water supplies, lack of access to medical treatment, for example immunisation against diseases such as measles.) Why do they think that most deaths in these countries seem to be in the earliest stages of the life cycle? **Answers:**

Cause of death	Uganda	Bolivia		Sweden
Malaria	128	25	5%	0
Tuberculosis (called 'TB')	89	50	10%	0
Measles	67	75	15%	0
Other causes	179	125	25%	27
TOTAL	500	500		500
Total died (A)	463	275		27
Total survived (B)	37	225		473

Cause of death of children in Bolivia

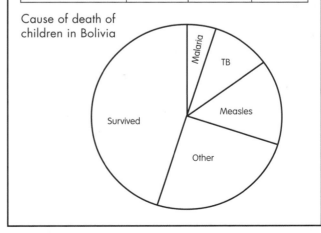

100 SCIENCE HOMEWORK ACTIVITIES ● YEAR 6

p98 Piggy-back tadpoles — LITERACY LINK

Learning objective
● To know how fertilisation occurs.

Lesson context
Look at the stages of a human baby's development, from fertilisation to birth (a timeline is available on page 21 of *100 Science Lessons: Year 6/Primary 7*). Ask the children to put the stages of development in order. Discuss the care provided by parents once a human baby has been born – do the children think that a human baby would survive without this parental care?

Setting the homework
This activity provides an opportunity to think about how other animals care for their young, and why parental care provides a better chance for offspring to survive. Explain that few animals provide as much care for their offspring as humans do. Introduce the strawberry dart frog, which lives in the Amazon in South America. Read through the worksheet, and encourage the children to find the Amazon on a map of the world.

Back at school
Discuss the children's answers. Define any words that may have proved difficult. Can the children think of other ways in which different animals care for their young? **Answers:** 1. The tadpole eats it as food!; 2. It is less likely that they will all be found by predators; 3. Most die because of predators, disease or poor weather; 4. The red colour is a warning to predators to stay away – the frog is protected by its poison!

p99 Reproduction crossword — SCIENCE PRACTICE

Learning objectives
● To know about the structure of the reproductive organs.
● To know how fertilisation occurs.
● To know how the growing foetus develops and how the baby is born.
● To know that changes in the body leading to sexual maturity begin at puberty.

Lesson context
Recap the structure and function of the male and female human reproductive systems with the children. Talk about the development of the baby from fertilisation through to birth, and the main changes that occur to the human adolescent during puberty. The children could bring in pictures of themselves as babies to reinforce the stages of their own development.

Setting the homework
You may wish to check in advance that parents are aware of the nature of topics covered in these lessons and that they are familiar with the school's sex education policy. This activity serves as a useful revision exercise to check that children are familiar with the terms covered in the lessons and that they understand their meanings.

You may want to run through the worksheet for those unfamiliar with the crossword format; you might write in the first letter of each clue on the grid for less able children.

Back at school
Go though the answers and clear up any areas of uncertainty that may have arisen. Are there any ideas they have thought of which can be credited even though they do not fit this particular crossword? Remember that there can be more than one correct name for an organ – for example, oviducts can also be known as fallopian tubes.

p100 Animal parents — FINDING OUT

Learning objective
● To know about the skills and care required in parenting.

Lesson context
Talk with the children about the special care parents need to provide for their young. If possible, invite a mother and baby to visit the class to talk about how they care for the young child. Remind the children that humans are mammals, and discuss the special features of this group – for example, the way that they provide food for their offspring (the mother makes milk in her mammary glands), the ability to keep a constant body temperature, the presence of hair or fur. Ask the children to think about the way other animal groups, such as fish or birds, may care for their young.

Setting the homework
Explain the sheet to the children. You might like to draw their attention to reference books available in the school library, and perhaps provide some photocopied background material for the children to take home to help them complete the questions.

Back at school
Share the group's answers, and look at a selection of examples. Have reference books or CD-ROMs on hand so the children can find out more about the five different groups. Ask the children to find out which parents lay the most eggs and why. Fish and amphibians lay many eggs, as most will be eaten by predators. Crocodiles (reptiles) provide some care for their young and so lay fewer eggs, but turtles, which abandon their young, lay many eggs. Birds usually feed their young and lay fewer eggs. Mammals tend to provide most care for their young, and have fewest young but with more surviving. **Answers:** fish – C; reptiles – A; mammals – D; birds – B; amphibians – E. The duck-billed platypus is an unusual mammal because it lays eggs, but still provides milk for its young!

UNIT 2 ANIMALS & PLANTS) VARIATION

p101 Coral reefs
FINDING OUT

Learning objective
● To know that living things can be arranged into groups according to observable features.

Lesson context
Ask the children to think about how living things might be sorted so we can tell them apart. Write a list on a board or flip chart of plants and animals, and ask the children how you might sort them (into *plants* and *animals*). Introduce the idea that scientists have divided living things up into 'kingdoms', and how the animal kingdom is further sub-divided, so animals can be classified as either *vertebrates* or *invertebrates*. Give the children lots of examples so they can try to do this themselves.

Setting the homework
Explain the sheet to the children. Choose one or two examples (without filling in any answers) to check that they understand the task. Try to provide reference materials to help those who may not have resources at home.

Back at school
Mark the sheet with the class, clarifying any uncertainties. Use a large map (perhaps on an OHP) to show the children the location of Australia and the Great Barrier Reef. If possible, show a video clip of a coral reef (the 'Coral reef' episode from the BBC's *The Blue Planet* series would be ideal). This work may link well with creative writing exercises about discovery/underwater adventure.
Answers: The sea slug, coral, starfish, urchin and crab are invertebrates; the angel fish, shark and turtle are vertebrates; seaweed is a plant. 1. Great Barrier Reef; 2. 'Crown of thorns' starfish (which are predators of coral), pesticides and fertilisers washed off the land, pollution, overfishing, climate change, sediments which settle on the reef and block out light and damage the coral polyps.

p102 Ancient animals
SCIENCE PRACTICE

Learning objective
● To know that arranging living things into groups based upon observable features can help in identifying unknown living things.

Lesson context
Make sure the children are familiar with the key external features of each of the five vertebrate groups by looking at pictures of a variety of creatures in each group and listing the differentiating similarities.

Setting the homework
Explain that though the animals are (probably) unfamiliar to them, the creatures – and their modern descendants – in each group share important features. Encourage the children to attempt the task using what they know about each group.

Back at school
Mark the sheets with the children. **Answers:**
1. Stegosaurus – reptile; Machairodus – mammal; Archaeopteryx – bird; Dinichthys – fish; Eryops – amphibian; 2. an animal with a backbone/spine.

p103 A guide to my family and friends
SCIENCE PRACTICE

Learning objective
● To use branching keys to identify an organism.

Lesson context
Recap on how we can classify organisms according to their characteristics. Ask the children how we might sort out a group of objects, and introduce the idea of keys. Look at an example of a sorting key (there is one on page 46 of *100 Science Lessons: Year 6/ Primary 7*). Ask groups of children to try to make their own key to help identify trees, for example by their leaves and twigs (using either pictures or real examples).

Setting the homework
Look at the way the key has been set out on the worksheet and brainstorm possible characteristics that could be used (for example, glasses/no glasses, male/female, beard/no beard). Remind the children not to use descriptions such as tall/short, as these are too vague – only physical features should be used. You might like to warn helpers that photographs may need to be brought in.

Back at school
Ask the children to swap their keys with others, who can test the keys to see if they work by seeing if they can match the names on the key with the pictures. Discuss any difficulties that arise and share ideas that worked well. The final keys will vary depending on the attributes used, but check that they have been logically structured.

p105 Variations
SCIENCE PRACTICE

Learning objective
● To know that animals of the same species vary.

Lesson context
Ask the children: *Can you roll your tongue?* Sort the class into two groups: those who can roll their tongue and those who cannot. Then re-sort the children by another, non-gender, criteria (for example, ask the children to stand in order of height). Discuss variations between individuals in the class, and talk about continuous (such as height) and discontinuous (such as blood group or tongue-rolling ability) variations.

Setting the homework
Read through the worksheet as a group, leaving the children to answer the questions at home.

Back at school
Ask the children to suggest answers. Prompt them: *Is soot such a problem today? Since the Clean Air Acts of the 19th century, which aimed to improve air quality, what do you think has happened to the number of pale peppered moths?* (They have increased in number.) *Why?* (They became better camouflaged against the trunks, which were lighter once the air became less polluted.) With whales, continuous variations are observed more easily than discontinuous variations. **Answers:** 1. differences; 2. wing colour; 3. They were better hidden from predators, and so survived to reproduce; 4. Variation means that some of the species are able to survive when their environment changes; 5. male/female; 6. pattern, fin size, length.

p106 Plants in space · LITERACY LINK

Learning objective
● To know that plants need light, water and warmth to grow well.

Lesson context
Revise what plants need for healthy growth. Let the children set up their own investigations into factors that affect plant growth, such as temperature or light, and encourage them to record their results over time. Emphasise fair testing.

Setting the homework
Even if you have not had time to do the investigation, this activity will provide an opportunity to revise the needs of plants. You may want to read through the worksheet as a group, leaving the children to answer the questions at home.

Back at school
See if you can organise a trip to a botanical garden or garden centre. @Bristol (0845, 345 1235; www.at-bristol.org.uk) has an excellent rainforest environment, as do the Botanical Gardens at Kew (020 8332 5000; www.rbgkew.org.uk) and the Birmingham Botanical Gardens (0121 454 1860; www.bham-bot-gdns.demon.co.uk), and there are plenty of other good examples around the UK. Use the trip to help the children to learn more about the growing requirements of plants and the conservation of important habitats. **Answers:** 1. oxygen; 2. photosynthesis; 3. any four from light, warmth, carbon dioxide, nutrients, water; 4. to provide food materials; 5. The plants could die, and in the dark plants use up oxygen. Relate these answers to the children's prior work on space: space is very dark and cold, so is this naturally a good plant environment?

p107 Growing tomatoes · NUMERACY LINK

Learning objective
● To know that plants need nutrients from the soil for healthy growth.

Lesson context
Consider the need for plants to have certain essential nutrients, including nitrates, phosphates, potassium, iron and magnesium. If possible, give the children the opportunity to examine healthy plants and compare them to ones lacking in certain nutrients to make comparisons.

Setting the homework
You might want to give the children one or two practice exercises in calculating averages so that they feel comfortable with the calculations needed for their homework. Finding the mean is a numeracy skill that Year 6/Primary 7 children should acquire and practise.

Back at school
Check through the children's answers. Ask individuals to model their calculations on the board so that the method is shared by the group. Discuss the need for *repeats* in experiments (averages help to show a general trend; one result alone could be a freak outcome). **Answers:** 1. A – 1kg, B – 2kg, C – 3kg, D – 0.5kg; 3. 20g, as this was the amount used in bag C that gave the greatest average yield; 4. This bag was a 'control' to show what would happen in an unaltered/normal situation with that type of soil; 5. The farmer would need to know how much soil (in kg) was in each bag to work out the concentration of nitrate (grams per kilogram) needed for the fields.

UNIT 3 THE ENVIRONMENT | THE LIVING WORLD

p108 Garden food webs · SCIENCE PRACTICE

Learning objective
● To know that food chains are used to describe feeding relationships in a habitat.

Lesson context
Using examples (such as grass > rabbit > fox), build up simple food chains introducing the key words associated with the topic, including *photosynthesis, energy producer, predator, prey* and *consumer*. Explain to the children that the initial energy for most food chains on Earth comes from the Sun, and is harnessed by plants (producers) at the start of the food chain. Ask the children to make their own food chains, and try to label them with the words *producer, consumer, herbivore, carnivore, predator* and *prey*. They must know that the arrows are important as they show the direction the energy is going in (arrows should show the flow of energy: from prey to predator).

Setting the homework
Explain the sheet to the children. You might like to read through the text together.

Back at school
Mark the sheets with the children. Ask: *What do the arrows represent?* (The flow of energy.) *Where do you think a mould/fungus should go in the food chain?* (Anywhere – all organisms decompose if they die!) **Answers:**
1. Completed food chain:

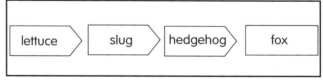

lettuce	slug	hedgehog	fox

2. lettuce; 3. from the Sun (it is used in photosynthesis to make food); 4. the hedgehog and fox; 5. There would be fewer foxes (fewer lettuces feed fewer slugs, which feed fewer hedgehogs, providing less food for foxes); 6. the flow of energy/direction the energy is going in.

p109 Adaptations for survival · SCIENCE PRACTICE

Learning objective
● To know that plants and animals have special features that help them survive in a habitat.

Lesson context
Introduce the idea that organisms (animals and plants) need to have special features to survive in their environment (to find food, find a mate, avoid predators, catch prey and so on). Draw a large outline of a shark and, as a group, brainstorm ways in which the shark is specially designed for its mode of life (powerful tail fin for swimming fast, sharp teeth for tearing flesh and so on). Label these on the diagram, and explain that these special features are called adaptations. Let pairs of children choose pictures of animals and plants from reference books and name the special adaptations of the organism they have chosen.

Setting the homework
You may wish to give children the opportunity to find out more about the examples by looking at resources (CD-ROMs or reference books) before taking the sheet home.

Back at school

Check through the children's work and see if anybody has learned any extra facts about the barn owl. If time permits, discuss why owls might be helpful to farmers (they prey on vermin) and what could be done to help barn owls survive, now that fewer barns exist today. A Venus fly-trap specimen would be a useful reference for the second question. In fertile soils, Venus fly-traps are overtaken by other plants which shade them: why could this be? (Venus fly-traps devote too much energy to growing traps instead of growing upwards, so they are soon covered up and die!) Explore adaptations in nature by looking at www.bbc.co.uk/nature, which features a quiz linking to the topic. **Answers:** 1. large eyes for spotting prey in dim light; downy wings for silent flight so they can catch prey by surprise; a hooked beak for tearing flesh; sharp talons for grasping prey; 2. traps made from leaves covered in hairs, which are sensitive to touch; overlapping spines on leaf trap bugs; digestive juices produced by leaf to dissolve the prey; traps and other green leaves make sugars using sunlight; roots anchor the plant to the marsh, helping it to take up water.

p110 Lichens and air pollution   NUMERACY LINK

Learning objective
● To use simple environmental survey techniques.

Lesson context
Show the children a square frame (a coat-hanger pulled into a square about 25cm × 25cm would be ideal) and explain that this is called a quadrat. Tell them that a quadrat is used by ecologists (scientists who study habitats) to study the plants growing in a habitat. Explain that it can be used to work out the area covered by each type of plant, and the number of different plant species found. Show how quadrats placed along a transect (often just a tape measure on the ground) can record the variation in plant species as an environment changes, for example from light to shade. If possible, go into a field or garden and give children the chance to carry out their own quadrat studies along a transect.

Setting the homework
Look at the children's quadrat surveys. Ask the children: *Which plant covers most of the ground? Was there a change in the types of plant found along the transect? How many different species were found?* Hand out the homework sheet and explain that it shows a similar type of study, but looks at only one type of organism (lichen).

Back at school
Check through the answers as a group. If time permits, take the group out to a local area (walls around the school, or perhaps a graveyard) to see if they can spot growing lichens. How many different lichen are found locally? What do the children think the local air quality is like?

There are many lichen study groups on the Internet; and the UK National Air Quality Information Archive can be found at www.aeat.co.uk, which gives detailed air quality readings for each region of the UK – find out more about the air quality in your area! **Answers:** 1. pollution (exhaust emissions) from vehicles; 2. A – this is found growing close to the road where exhaust emissions are more concentrated; 3. C – this is sensitive to emissions, and seems to grow well only in cleaner air; 4. The prevailing wind blows exhaust emissions in an easterly direction, so the pollution is more concentrated here.

p111 The soil where I live   FINDING OUT

Learning objective
● To know that different soils can be compared.

Lesson context
Provide groups of children with samples of soil. Ask them to put the soil into a glass of water and observe how the soil particles settle after shaking. Use this to illustrate how soil is made up of rock particles (of varying sizes) and humus (which usually floats on the top). Explain that humus is made up from animal and plant remains, and that it is an important feature of soil – as are the water content and air spaces between the particles.

Setting the homework
Tell the children that they will be looking closely at the soil near their home. For those without a garden, you may want to provide a sample of soil from the school grounds in a sealed container to take home. Remind the children that they should not handle soil if they have any open cuts on their hands, and that they should work with their helper when carrying out this investigation.

Back at school
See if children living in the same area have classified their soil similarly (there will of course be very local differences). Which type of soil seems most common for the area? You could invite a local arable farmer to talk to the class about the local soil – which type of crops grow best in the soil? How does the farmer maintain the quality of the soil? Try setting up a wormery to find out more about worms. **Answers:** 1. Answers will vary; 2. Loam soils are a good mix of clay and sand – they allow water to drain, but hold enough back to support plants after the rain has passed; 3. Worms are slim and muscular, and pull themselves forward using bristles that grip the soil. They pass soil particles through their body as they move, extracting nutrients. Their trails help water to drain and aerate the soil, which they keep mixed. By pulling leaves into their burrows they help add nutrients to the soil.

p112 Preventing decay   LITERACY LINK

Learning objective
● To know that the decay caused by micro-organisms is useful in the environment.

Lesson context
Talk about what happens to the bodies of animals when they die. Develop the idea that tiny microbes feed on dead material, and introduce the term *decomposers* (mainly bacteria and fungi). Explain that decomposers help to release vital nutrients back into the soil so they can be used again, and ask the children to think why this is important. Explain that these microbes, like most other living things, need certain conditions to live, grow and reproduce, and that this needs to be considered when decay is wanted (for example, in a compost heap) and in situations where the aim is to prevent decay (for example, the need to keep food fresh).

Setting the homework
Having looked at the role of microbes in decomposition, explain the worksheet to the children. Remind them that decay is caused by living things, which might help them answer the questions.

Back at school
Check through the answers to see what the children have

understood about what microbes need in order to grow and reproduce. You may want to look at preservation techniques through the ages, such as drying, salting, smoking, freezing, sterilising (by heating) and, more recently, irradiating. Which methods would have been used a thousand years ago? Which are still common today? **Answers:** 1. Microbes were inactive because of the cold; 2. Chemicals in the tar have acted as preservatives, killing or inhibiting microbes; 3. Removing moist tissues – microbes love water! 4. Many microbes, like most living things, need oxygen to survive; 5. moist, warm, with air, without toxic chemicals.

p113 Useful microbes — SCIENCE PRACTICE

Learning objective
● To know that micro-organisms are used in food production.

Lesson context
Bring in a number of everyday examples that show how microbes can be put to good use. Bread, beer, wine, cheese and yoghurt all need the help of microbes at some stage of their manufacture; pea plants harbour useful bacteria in their roots, and these make nitrates that enrich the soil; microbes help to break down sewage at water treatment works. Give children the opportunity to access resources to learn more about each of these processes.

Setting the homework
Explain the sheet to the children. Say that they can use reference materials to help them learn more about the processes invoved in manufacturing each food.

Back at school
If possible, display a range of the food products shown on the sheet to reinforce how microbes contribute to our everyday life. Interestingly, many types of plant cannot survive without friendly fungi living in their roots, helping them to absorb nutrients. We even have bacteria living in our gut which help us to make certain vitamins! **Answers**: 1. – D; 2. – B; 3. – F; 4. – C; 5. – A; 6. – E.

UNIT 4 MATERIALS REVERSIBLE AND NON-REVERSIBLE CHANGES

p114 Soluble or insoluble? — SCIENCE TO SHARE

Learning objectives
● To know that some materials dissolve in water.
● To know that solids that do not dissolve in a liquid can be separated from it by filtering.
● To be able to separate an insoluble solid from a liquid by filtering.

Lesson context
Remind the children what *dissolving, insoluble* and *soluble* mean. Show that sand does not dissolve in water, but salt and sugar do, then explain that insoluble solids (such as the sand) can be separated from water by filtration. Demonstrate the process of separating sand from water, using filter paper and a funnel, then show that you cannot separate salt from water by the same process.

Setting the homework
As this activity requires specific materials, you might want to warn helpers in advance of this activity. Read through

the sheet together, making sure the children understand what they have to do. Stress that they might not have all the foodstuffs listed, and that this doesn't matter – they need not go out and buy things if they don't have them. Emphasise the health and safety point of using only tepid water.

Back at school
Discuss the children's experiences and results. Did anybody try any other foodstuffs? What happens when oil and water are mixed? **Answers:** 1. Flour should be insoluble in water, and all the others should be soluble. 2. filtration using filter paper in a funnel held over a beaker/jar. (Some children might suggest using home-made apparatus such as kitchen roll or a muslin bag rather than filter paper, and a tea strainer instead of a filter funnel.)

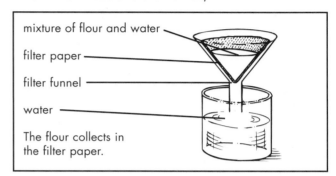

mixture of flour and water

filter paper

filter funnel

water

The flour collects in the filter paper.

p115 Ice and water — SCIENCE PRACTICE

Learning objective
● To know that when a gas is cooled it becomes a liquid and that this process is called condensing.

Lesson context
Recap on previous work (see Unit 4 of *100 Science Lessons: Year 4/Primary 5*), which looked at solids changing into liquids (melting) and vice versa (freezing). Reinforce the idea of evaporation, then go on to explain what is meant by the term *condensing*, giving everyday examples (such as water on a kitchen window when someone is cooking).

Setting the homework
Read through the worksheet with the children. You may need to prompt the children's thinking: the ice cools down the water, which in turn will make the glass cold. The glass is surrounded by air, which is a gas. What will happen when the air touches the cold glass? (Condensation.)

Back at school
Make a copy of the graph on an OHT, and show this to the children. Go through each of the questions on the worksheet and check the children's answers. You could ask additional questions such as: *If you didn't put ice cubes in the glass of water, would you still have seen the droplets on the outside of the glass?* (This would depend on how cold the water was.) *Would the temperature of the water fall if you didn't put in an ice cube?* (Not unless the room became very, very cold.) **Answers:** 1. condensation/condensing; 2. from the air/atmosphere, water vapour; 3. melt or reduce in size or turn to liquid/water; 4a. 11°C, b. This is the temperature of the water in the taps or pipes. Accept that this is the temperature of the room/surroundings, but this is not necessarily correct because room temperature is usually about 20°C; 5. 4°C; 6a. lower, b. The ice cooled it down/the ice melted to form cold water which cooled the tap water down; 7. A; 8. greater than 18°C but less than 20°C.

p116 & 117 The dissolving sugar experiment 1 and 2 — NUMERACY LINK

Learning objectives
● To know how the temperature of water affects the speed of dissolving.
● To be able to interpret results.

Lesson context
Ask the children to carry out an experiment to show how the temperature of water affects the speed at which sugar and/or salt dissolves. Use four different temperatures of water: fridge (about 4°C); room temperature (about 20°C); and two samples of heated water (30°C and 40°C, but no hotter to prevent scalding). Focus on predicting the outcome of the experiment, and emphasise the need for a fair test – for example, using the same amount of water and sugar. Recap how to tabulate results and draw line graphs; you might like to introduce more able children to judging a line of best fit, as used on the second homework sheet.

Setting the homework
These two activities focus on different aspects of the same experiment. You could set them together, or on consecutive nights. The first sheet looks at fair testing and carrying out the experiment safely; the second concentrates on interpreting and plotting results as a graph. Read through each sheet with the children, making sure that they understand the questions.

Back at school
Discuss the children's answers. Give the children the opportunity to volunteer their answers. **Answers:** Dissolving sugar experiment (1): 1. 135 seconds; 2. 23 seconds; 3. The sugar would dissolve faster at 50°C than it did at 40°C; 4. could cause scalding; 5. used the same amount of water and sugar and same sized bowls. Dissolving sugar experiment (2): On graph, look for clear and accurate points plotted, line of best fit, neatness and care; 1. 56 or 58°C; 2. 12 or 13°C.

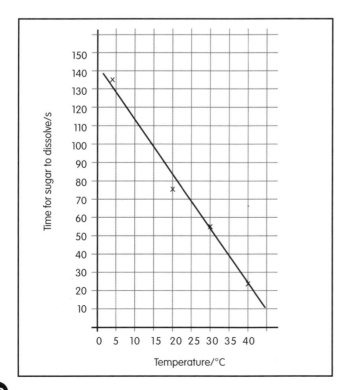

p118 Mixing materials — SCIENCE PRACTICE

Learning objective
● To understand that changes sometimes happen when materials are mixed together, and that these changes cannot be reversed easily.

Lesson context
Briefly recap earlier work on mixtures and separating mixtures. Explain the difference between a reversible and an irreversible change (ask the children to think how they would get salt back from a salt solution). In groups, ask the children to observe what happens when they mix plaster of Paris with water and then allow it to set; or what happens when they add Andrew's Liver Salts to water. Explain that the bubbles being formed by the liver salts mean that a gas is being produced – do the children think they will be able to get the original materials back?

Setting the homework
Explain that to answer the questions, the children will need to use the information in the table on the worksheet.

Back at school
Ask the children to discuss their answers in small groups. One person from each group can then feed back the answers to the class, where the children can discuss the ideas if necessary. **Answers:** 1. water, oil, vinegar; 2. bag A; 3. A gas was given off (carbon dioxide); 4. bag C.

p119 Burning — SCIENCE PRACTICE

Learning objective
● To know that burning brings about changes that are irreversible.

Lesson context
Demonstrate burning a wooden splint, allowing the children to observe what happens – note that flames, smoke and heat are given off, and that there is a new end-product (ash) that is totally different to the splint you started with. Explain that burning is an *irreversible* change because the ash, smoke and flames cannot be turned back into the wooden splint.
⚠ When demonstrating burning, it is essential to consider the risks and necessary health and safety precautions. The burning material could cause other things to catch fire; have a fire blanket and a bucket of sand close by. Be aware of the immediate first aid measures for the treatment of burns (for example, CLEAPPSS' Model Safety Policy).

Setting the homework
Read through the worksheet, and make sure that the children know what they have to do. Remind them that they should not try to recreate the experiment at home.

Back at school
Discuss the children's responses, and recap the nature of reversible and irreversible changes. **Answers:** 1. flames, smoke, grey ash left (all of these should be mentioned for a correct answer); 2. chocolate and wax; 3. The mass of the ash left would be less than that of the wooden stick (2g); 4. irreversible change.

(p120) Rusting

Learning objectives
- To know that only iron and steel rust.
- To know that oxygen and water are needed for rusting.
- To know that rusting is an irreversible change.

Lesson context
Discuss rusting in our everyday lives and what might cause it (iron, water and oxygen are all required to produce iron oxide – otherwise known as rust). Set up an investigation to find out what causes rusting and which materials (say, from a selection of iron, steel, copper and aluminium) actually rust. Remember to consider how to make the investigation fair, and ask the children how they could record their observations over a period of time.

Setting the homework
This activity is best set after the children have completed their investigation and the class results have been looked at. Remind the children of the experiment, and go over the worksheet so they understand what they have to do.

Back at school
Discuss the answers with the children. Can any of them think of objects that will rust other than those in the picture? **Answers:** 1. iron and steel; 2. water and air/oxygen; 3. bike, wheel, washing line pole, railings, barbecue, nails, pram, bolts on garden gate, lawn roller; 4. bike and wheel – oil or grease metal parts, especially the bike's chain, and keep them dry where possible; pole, bolts and railings – paint them; barbecue – cover it up or keep it out of the rain; nails, pram and lawn roller – keep them dry. 5. Sea water contains salt, so if sea water gets onto the car or bike it will contain salt which will speed up rusting.

(p121) Generating electricity

Learning objective
- To know that electricity is made from non-renewable fuels.

Lesson context
Talk about things the children found at home that require electricity to work. Move on to talk with the children about how electricity is made (for example, coal is burned to heat up water, which produces steam that turns turbines to produce electricity). If you have secondary resources (posters or videos), show the children this process in action.

Ask the children to draw diagrams that illustrate the different stages of electricity production (alternatively, provide a sequence of pictures and ask the children to put them in the correct order – page 120 of *100 Science Lessons: Year 6/Primary 7* has some suitable pictures).

Setting the homework
Introduce the children to the scientist Michael Faraday, and his contribution to discovering electricity. Explain that you want them to write a newspaper article about Michael Faraday's work. How much detail you ask for will depend on the children's ability, but the questions on the worksheet will provide some guidance for them. You might like to suggest some sources of information for the children to turn to when researching their articles (the school or local library, for example).

Back at school
Ask some children to read their articles out to the class. The articles could be used as part of a display about electiricity.

(p122) Traffic lights

Learning objective
- To know that the number of batteries and bulbs in a circuit can affect the brightness of the bulb(s).

Lesson context
Using a selection of batteries, bulbs and wires, ask the children to build electrical circuits and observe the effect that changing the number of batteries has on the brightness of the bulb(s), and other components, in the circuit. Encourage them to try one battery with one, then two, bulbs, then two batteries with one and two bulbs – remember to change only one component at a time.

Setting the homework
For this activity, the children need to be aware of how switches control the electricity passing through a point in a circuit. The children should be able to follow the paths that the electricity will take around the circuit on the homework sheet, even though this is a parallel circuit, which they may not have encountered before. You could explain that a parallel circuit allows each bulb to be operated independently.

Back at school
Check the children's answers together. **Answers:** 1a. switch 1 only, b. switches 1 and 2, c. switch 3 only, d. switch 2 only, e. switch 1 only; 2. All the bulbs are connected independently to the battery.

(p123) Matching up

Learning objective
- To know that electrical circuits and components can be represented by conventional symbols.

Lesson context
Introduce the children to the conventional circuit symbols shown on the worksheet. Discuss why those symbols are used rather than drawings of the components. (Because these are easy for everybody to understand.) Draw some simple circuit diagrams on a board or flip chart (there are some examples on page 127 of *100 Science Lessons: Year 6/Primary 7*), and ask the children to interpret these diagrams to build electrical circuits. Start with familiar battery–bulb combinations.

Setting the homework
Make sure that the children are all familiar with the symbols used on the worksheet. Explain the task to the children, and emphasise that when drawing circuits, batteries are often grouped together and wires are drawn as straight lines, which results in the circuit diagram being rectangular.

Back at school

Check the children's answers and go over any common problems or misconceptions. **Answers:** 1. Correct symbols:

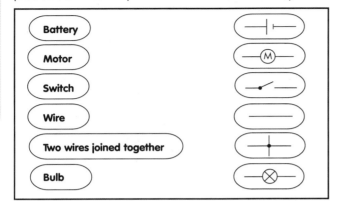

Battery	—⊢—
Motor	—Ⓜ—
Switch	—•⁄—
Wire	
Two wires joined together	—⊢—
Bulb	—⊗—

2. Circuit diagrams:

a.   b.

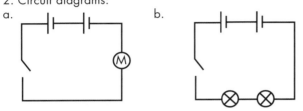

p124 The combination is...   SCIENCE PRACTICE

Learning objective
● To know that switches can be placed between parts of a circuit to provide alternative routes for current to pass along.

Lesson context
Introduce the children to different types of switches in circuits. Build circuits containing switches that need to be pressed in combinations or where a choice of switches will result in the same operation. For example:

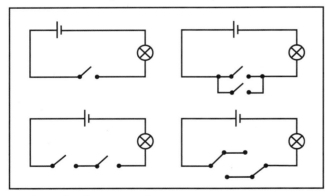

Setting the homework
Explain the sheet to the children, and point out that this activity shows that switches can control more than one device. The children need to be able to follow the path of the electricity around a circuit, so you might like to suggest following the path of the electricity with a pencil line that can be rubbed out when another path has to be followed.

Back at school
Discuss how, and why, different bank workers were able to open the doors leading to different areas of the bank. Make sure the children know that no one person would be able to open and enter the most restricted area – the safe – alone. **Answers:** 1. two; 2. Manager; 3a. Manager and Assistant Manager or Manager and Cashier, b. All three are needed, c. any of the three alone.

p125 The safety code   LITERACY LINK

Learning objective
● To recognise the dangers associated with electricity.

Lesson context
Show the children a short circuit. Explain how it can occur and its effects, particularly how dangerous it can be – when electricity 'shorts' its proper path, too much current may travel through too thin a wire, causing overheating and possibly resulting in a fire. Overloading a socket is a 'common' unsafe electrical practice in the home. As in a short circuit, this puts too much current into one part of the (house) circuit, which is not designed to cope with it. Using more individual sockets spreads the load on the house ring main, which can carry such a current safely.

Setting the homework
Explain the sheet to the children. Point out that the rules should be written in numbered or bullet form rather than as continuous writing.

Back at school
Compare the children's lists and formats of the rules. Talk about how to use electricity safely.

p126 Wired up   NUMERACY LINK

Learning objective
● To know that the amount of electricity flowing in a circuit is related to the total resistance.

Lesson context
Investigate how increasing the length of a nichrome wire in a circuit can affect the speed of a motor. Discuss how the wire causes resistance – a property of wire that determines the ease with which electricity will flow through it. Increasing the resistance makes it harder for the electricity to flow, and reduces the current, slowing the motor.

Setting the homework
Recap the investigation into resistance, and make sure the children understand the worksheet. The graph plots the resistance of the wire for different lengths, and should produce a straight line. Remind the children of the features of a correctly drawn graph. (You may like to draw in the axes for less able children, and even add the units. The graph paper allows for a scale of 1:2 to be used.)

Back at school
Check the graph has a title, labels and was drawn with a ruler. **Answers:** Completed graph:

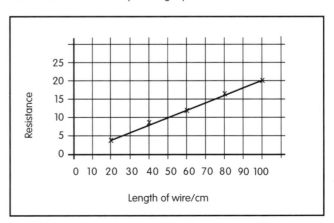

Missing words (or similar phrases): larger, 5, 10, twice as big.

p127 Many hands make light work
SCIENCE TO SHARE

Learning objective
● To know about the construction of electrical circuits in real life.

Lesson context
Think about real-life electrical circuits, for example those in the home, and use secondary sources to find out about methods of saving energy.

Setting the homework
As the children near the end of this topic, and the last time they will study electricity in this phase, this homework draws together real circuits, conductors and insulators and electrical safety. Emphasise the importance of having an adult to help and supervise them while doing this homework.

Back at school
Look at the children's ideas. They may have identified the glass bulb as transparent and an insulator, and the screw/bayonet fitting, terminals and filament as metal conductors. Make sure they realise the electricity never has to pass through the glass of the bulb.

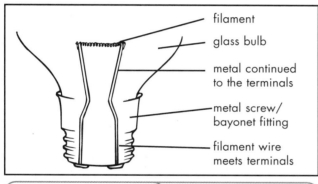

filament
glass bulb
metal continued to the terminals
metal screw/bayonet fitting
filament wire meets terminals

UNIT 6 FORCES & MOTION — FORCES AND ACTION

128 Your next game is...
SCIENCE PRACTICE

Learning objective
● To know that a force exists between two magnets, and between magnets and magnetic materials.

Lesson context
Remind the children how magnets behave around each other – that they attract or repel other magnets depending on the way the poles are positioned with respect to each other. Let the children measure the force between a magnet and a steel paper clip by trying to pull the paper clip away from the magnet using a newton meter.

Setting the homework
Explain the worksheet to the children. Make sure they are aware what properties of a material explain whether it is magnetic or not, and how they can test for this.

Back at school
Ask some children to share their answers with the class.
Answers: Elizabeth must identify the magnets by finding which things in the box are attracted with one pair of their ends and repelled with their other ends. Once she has identified the magnets, she can use them to test the other things to see if they are magnetic. The steel can and iron bar are magnetic; the aluminium foil and copper disc are not.

p129 Newton's diary
LITERACY LINK

Learning objective
● To know that the force of gravity is responsible for the weight of an object.

Lesson context
Using a newton meter and a set of scales, ask the children to compare the mass (in g or kg) and the weight (N) of a variety of objects. Help them to notice the relationship between the values of the objects – that the weight in newtons is approximately ten times the weight in kilograms. Talk about gravity as the force on an object that results in its weight.

Setting the homework
This is a two-part task that would be ideal for extended project work, perhaps over the holidays, or to be split between home and school time. Introduce the children to the work of Sir Isaac Newton. Explain that he is the scientist who 'discovered' gravity, but that he did a lot more for science than just this. Tell the children that they should do some research about Newton's other famous discoveries. Suggest where they could find information (if children don't have resources at home, you could ask them to look in the school library). Remind the children not to copy passages of information, but to identify the relevant parts of their research and report this in a format appropriate to their intended audience. You may like to assign different questions to different groups to ensure a variety of research and creative writing.

Back at school
Let the children compare their writing. The work could be displayed, or passages enlarged from it to enhance a display about Sir Isaac Newton (and other scientists).

p130 Bungee!
NUMERACY LINK

Learning objective
● To know that when the force stretching an elastic band is increased, its length increases in proportion.

Lesson context
Ask the children to conduct an experiment to find out how the force on an elastic band affects its length. Start with an unstretched elastic band, then use a force meter to stretch the band by an increasing amount, recording the change in length and plotting the results as a graph. The graph should show a proportional relationship between the length of the band and the force applied.

Setting the homework
Remind the children of what they found when stretching the elastic band, and explain the worksheet to the children.

Back at school
Look at the shape of the graph together, noting the proportional stretch in the rope as the greater weight/force is applied. Point out that eventually the rope becomes overstretched and will break – this is the rope's *breaking strain*, and only people below a certain weight would be allowed to use the bungee rope. This maximum weight would be carefully calculated to ensure everyone is safe.
Answers: 1. 50m; 2. 70m; 3. 20m; 4. 1000N; 5. just over 118m; 6. (safely) 1200N – about 18st.

p131 Floating metal — SCIENCE TO SHARE

Learning objective
- To know that when an object floats, the upthrust acting on it is equal to the force of gravity and acting in the opposite direction.

Lesson context
Carry out an activity to measure the forces on objects that float and those that sink. The difference between the weight (in N) of an object in air and the weight (in N) of the same object immersed (not floating!) in water is known as *upthrust*. If this experiment is repeated with objects that float and ones that sink, you will show that gravity pulling down equals the upthrust pushing up from the water for objects that float.

Setting the homework
The children will be aware (or you can show them) that solid pieces of metal usually sink. They may also know of pond creatures that float on the surface of a pond – these use *surface tension* to stay on the surface. It is worth discussing this invisible 'skin' on the surface of water, which can support an object such as a pond skater (or a needle, the piece of solid metal in the activity). Explain that the 'skin' comes from the way that the particles in the water are connected to each other. Point out to the children that they will need to carry out the experiment at home with great care and patience.

Back at school
Discuss the children's experiences. Explain that the needle floats because the upward force of the surface tension is greater than the downward force of gravity. The needle should sink when the washing-up liquid is added, because this destroys the forces that cause surface tension when it mixes with the water.

p132 The daredevil — LITERACY LINK

Learning objective
- To be able to describe how air resistance can slow down a falling (moving) object.

Lesson context
Using identical weights and a range of home-made parachutes of different sizes, explore how the area of a parachute affects the time it takes for an object to hit the ground, and therefore its speed. Emphasise fair testing – using the same weight and dropping from the same height. Recap on the ideas of air resistance and how this acts on the parachutes to slow them down.

Setting the homework
Read through the description of Stuart the parachutist's activities with less able readers. Make sure the children understand that they must use the information in the passage *and* their work in lessons to answer the series of questions.

Back at school
Go through the answers together. **Answers:** 1. The truck has more air resistance than the sports car; 2. The aeroplane is not supported by the wheels any more, but by the air passing under the wings. More able children might suggest that the *lift* created by the wings moving through the air equals gravity pulling the plane down – it *floats* in the air; 3. There is a greater force downwards from his weight than from air resistance upwards; 4. The air resistance increases due to the greater surface area of the parachute; 5. The smoke would be blown across the field in the direction of the wind; 6. Stuart should make sure he bends his knees on landing to absorb some of the shock; 7. He will feel the air resistance in his face and hair.

p133 The gymnastics display team — SCIENCE PRACTICE

Learning objective
- To be able to explain how objects stay at rest or move by considering the forces acting on them.

Lesson context
Look at the forces acting on a variety of different objects, and demonstrate how pairs of forces can balance out to hold an object in place (such as the support of a table pushing up and the force of gravity pulling down when a book is placed on the table). Ask a child to hold a tennis ball on a string in the air – ask them to point out where the forces are acting (gravity pulling down and the string pulling up).

Setting the homework
The homework activity covers different forces from those looked at in class. Make sure the children are able to name a range of forces (*gravity, friction, air resistance* and so on), that they know that forces have direction and how forces act to make an object move or stay in place.

Back at school
Go through the answers together. You may like to consider some of these moves in a PE lesson if you have any 'expert' gymnasts in the class – can they feel forces acting on them if, for example, they wobble? **Answers:** 1. stretching/ pulling forces – the stretched arms hanging off the rings and bars, the gymnast on the floor in a stretched position; compressing/pushing forces – the gymnast on the floor, the weight of the gymnast doing a handstand on the beam; 2. gravity or weight; 3. The bar is bending; 4. The weight of the gymnast is being taken in the tension of the ropes and the gymnast's arms.

UNIT 7 LIGHT & SOUND — LIGHT AND SOUND AROUND US

p134 Staying in the shade — SCIENCE PRACTICE

Learning objective
- To revise how light travelling from a source can be blocked by an opaque object, making a shadow.

Lesson context
Give small groups of children a simple cut-out shape and a lamp. Ask them to investigate the relationship between the position of an object and a light source, and how this affects the size of shadow cast. Make sure the children understand and can use the terms *source*, *opaque* and *shadow* correctly when they develop a plan for their investigation, and in their analysis of results.

Setting the homework
Make sure the children understand that they have to use their knowledge of light and shadows in order to complete some of the questions.

Back at school
Check the children's answers. **Answers:** 1. warm/hot; 2. to keep cool, or to keep out of the sun; 3. C; 4. The tree is opaque, so light cannot pass through it; 5. The shadow would become hazy or would disappear.

p135 True or false?   SCIENCE PRACTICE

Learning objective
● To know that non-luminous objects are seen because light scattered from them enters the eye.

Lesson context
Darken the classroom, and show the children that in order to see a non-luminous object better in a dimly lit room, we need to shine light from a light source (such as a torch) at the object so the light is scattered (reflected) from the object and into our eyes. Draw a diagram showing how light from a source (such as the Sun), strikes an object and scatters into our eyes so we can see it.

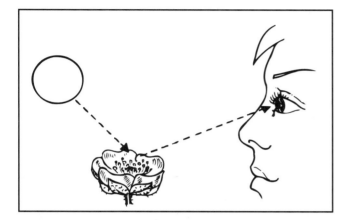

Setting the homework
The statements on the worksheet aim to address some common misconceptions about how we see things. Remind the children that many adults don't understand how we see things either – so they might like to test their helper as well! You could read through the sentences with less able children before sending it home.

Back at school
Mark the worksheet with the children. **Answers:** No, it's not a mistake – all the answers are true! Since the concepts themselves can be tricky, this activity reinforces the truths and not the misconceptions.

p136 Seeing things differently   SCIENCE TO SHARE

Learning objectives
● To know that mirrors can be used to change the direction in which light is travelling.
● To understand that shiny surfaces can be used for mirrors, but dull ones cannot.

Lesson context
Compare the reflective qualities of different surfaces (such as mirrors, aluminium foil, painted surfaces). Ask the children to compare the surfaces – can they see their reflection in each surface? Encourage them to explain their findings in relation to the properties of the materials. With more able children you may like to investigate the relationship between the rays of light striking a mirror and those coming off the mirror: can this help us to see round corners?

Setting the homework
Ask the children to look around the classroom to see what reflective surfaces they can find. Explain that this worksheet shows how curving a mirror (such as a spoon) can produce useful and also fun effects.

Back at school
Display the children's drawings alongside some examples of curved mirrors. IKEA produce some interestingly shaped mirrors that could be used to enhance a display.

p137 What a view!   LITERACY LINK

Learning objective
● To know that differences in reflectivity and shadow formation affect how we see our surroundings.

Lesson context
Explore the classroom together, noting how and where shadows are formed, and what effect they have on the way we see things. Ask the children: *Do the shadows change at different times of the day?* Move on to map the school environment, looking at where shadows form and reflections occur.

Setting the homework
The children may need to carry out some preparatory work on descriptive language and how language can be used to conjure up emotions before this activity. If any children have trouble finding pictures, you might like to provide some magazines or books for them to take a picture from. Emphasise description of the light and shadow, not the physical details, in the scene.

Back at school
Look at some of the children's pictures, and ask some to read out their descriptions. Display the descriptions along with the pictures. In art time, explore drawing the effect of light on a place – a key theme of the Impressionists' work. Teachers in England could link this activity to QCA Art Unit 6C: 'A sense of place'.

p138 On the way   SCIENCE PRACTICE

Learning objectives
● To reinforce the idea that parts vibrate to produce sound.
● To know how pitch and loudness can be changed.

Lesson context
Ask the children to bring in (or you could provide a set of) different musical instruments to play. Investigate how the sound is produced by each. Explain that all instruments produce sound by vibrating the air around them. Explore how the sounds can be changed.

Setting the homework
Ask the children to think about the different sounds they might hear in the environment on a typical day. Read through the passage together. Make sure all children are aware of the relationship between vibrations and sound, and they realise that the larger the vibration is, the louder the sound will be.

Back at school
Re-read the passage together, and ask the children to point out when sounds are mentioned. **Answers:** 1. people – Mum, bus driver, friends, class, teacher; 2. objects – doors, cars, shower, spoon in the bowl, door, engine, horn, bell, hinges; 3. a source; 4. It vibrates.

p139 The speed of sound
NUMERACY LINK

Learning objective
● To reinforce that sound can travel through solids, liquids and gases.

Lesson context
Discuss how sounds in school, and the wider environment, can be heard through walls and over large distances (such as across the playground), showing that sound can travel through both solids and gases. Look at secondary sources and set up demonstrations to observe how sound can pass through each of the different states of matter (for example, how you can hear when underwater in the bath or a swimming pool – this can be demonstrated using a stethoscope in a tank of water – and using string telephones to hear through a solid).

Setting the homework
Using charts with axes marked in multiples and using data to solve problems are important skills to reinforce in Years 5 and 6/Primary 6–7. Remind the children about the need to choose and use a scale for the vertical axis on the bar chart, with evenly spaced numbers and neat, separate vertical columns. For ease of comparison, you could suggest that more able children sort out and plot, for example, all the solids, then all the liquids and then the gases next to each other.

Back at school
Look at the graphs. Check that the graph has a title, appropriate labelling, and that the children have chosen an appropriate scale (1cm per 500m/s or 1000m/s). Note that the speed of sound through a material can vary with temperature and the size of the sample, which explains the variance in the results on the worksheet.

Answers:

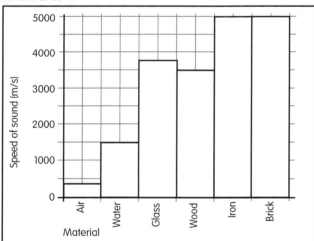

UNIT 8 EARTH & BEYOND | THE SOLAR SYSTEM

p140 The Sun, Moon and the Earth
SCIENCE PRACTICE

Learning objectives
● To know that the Earth spins as it goes around the Sun, and that the Moon travels with the Earth and in orbit around it.
● To reinforce the relative sizes of the Sun, Moon and Earth.

Lesson context
Using video, models, pictures and discussion, show the children the relative sizes of the Sun, Moon and Earth. A beach ball, a pea and a poppy seed would be good choices for models – the Earth is about four times bigger than the Moon; the Sun is about 400 times bigger than the Moon and about 109 times the size of the Earth.

Demonstrate the distance between the Sun, Moon and Earth. Using the same scale (which is roughly 1mm:10000km), the Moon (poppy seed) will be about 4cm from the Earth (pea), and the Sun (beach ball) about 15m from the Earth.

Remind the children that the Earth spins on its axis as it goes around the Sun: that the Earth rotates on its axis every 24 hours, and orbits the Sun every 365¼ days. Remind the children also that the Moon travels with the Earth and in orbit around it. It takes 28 days for the Moon to orbit the Earth.

Setting the homework
Read through the worksheet with the children, and make sure they know what they have to do.

Back at school
Discuss the answers with the children. **Answers:**
1. The direction from which the Sun is shining onto the Earth:

2. on its axis and around the Sun; 3. reflected, Sun; 4. 28 days; 5. 24 hours, darkness, Sun.

p141 An eclipse of the Sun
LITERACY LINK

Learning objective
● To know how a solar eclipse occurs.

Lesson context
Describe, using a model set up as shown below, how a solar eclipse occurs and what we see.

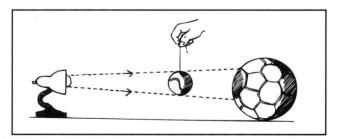

Talk about the solar eclipse that occurred in the summer of 1999. Images are available on the Internet at www.eclipse.org.uk.

Setting the homework
Remind the children of what happens during a solar eclipse. Explain the worksheet to the children. You might like to read through the passage with less confident children.

Back at school
Read through the passage together and fill in the blanks. **Answers:** solar, Earth, Moon, umbra, bigger, further.

p142 The life and work of Galileo Galilei — FINDING OUT

Learning objectives
● To be able to describe the surface of the Moon.
● To know how the craters on the Moon were formed.

Lesson context
Show the children images of the surface of the Moon (pictures are available from www.nasa.gov). Ask the children to look at the pictures and describe them, then explore information about the scientists who first saw these things. Point out that the surface of the Moon is covered with craters. Explain how these craters are formed (demonstrate by dropping different-sized pebbles into a tray of sand or flour) and why they remain on the Moon's surface because of the lack of atmosphere.

Setting the homework
Introduce the children to the scientist Galileo Galilei. Explain a little about his scientific discoveries, and tell the children that they should write a short report on some of his famous discoveries. Explain how much detail is required (this, of course, will vary with different ability groups). The following are the key events in Galileo's life that you might like the children to focus on: born 15 February 1564; makes his first telescope – 1609; 'The Starry Messenger' is published – 1610; a dialogue concerning the two chief world systems is published – 1631; sentenced to prison for heresy – 1633; died 8 January 1642.

Talk about where they might find information, such as the town or school library (for example, *Heavens Above: The Story of Galileo Galilei* by Kenneth Ireland, Super Scientist series, Macdonald Young Books, 1997) or websites (The Institute and Museum of the History of Science of Florence, Italy; http://galileo.imss.fi.it/museo/4/index.html).

Back at school
Ask the children to pool together their information, working in small groups. Help them to role-play aspects of Galileo's life, such as when Paolo Sarpi told Galileo about Hans Lippershey making a telescope, so Galileo rushed home from his holiday to make his own telescope, thereby not allowing Hans Lippershay to get his invention noticed first.

Put the children's dramas together and let individuals read out what they have found out. Use the reports as the basis for a complementary display, or for an assembly presentation linked to work on space, scientific enquiry or famous historical characters.

p143 The Solar System — SCIENCE PRACTICE

Learning objectives
● To know what makes up our Solar System.
● To know the order of the planets in our Solar System.
● To know which planets have moons.

Lesson context
Introduce the children to the planets in our Solar System through the use of videos, websites, CD-ROMs, books and posters. Encourage them to find out the order of the planets, what the planets look like, which planets have moons, and what the asteroid belt is made of. Be aware that you may have a space 'expert' in the class, so use him or her as a resource to support the class.

Setting the homework
Depending on the focus of your work in class, some of this sheet may be revision, while other aspects will need fresh research. Explain the sheet to the children, and make sure they understand what a mnemonic is to help them remember the order of the planets.

Back at school
Mark the questions with the children, by allowing the children to volunteer their own answers first. Share some of the children's favourite rhymes from question 4. **Answers:** 1. Solar System; 2. planet, Solar System, galaxy, Universe; 3. The two missing labels are Mercury and Venus; 4. A good rhyme to remember the order of the planets is *My Very Easy Method Just Speeds Up Naming Planets*, although the children may suggest more creative alternatives.

p144 How far from the Sun? — NUMERACY LINK

Learning objective
● To know the order of the planets in our Solar System.

Lesson context
Solving problems by extracting and interpreting data in graphs is a key maths skill that is very relevant to science. Use music (such as Holst's *The Planets Suite*) to provide an atmospheric introduction, and to stimulate the children's imaginations as they use reference sources to find out more about the Solar System, including what the planets look like and the distance of each planet from the Sun. You may like to make a model of the Solar System. If you use a cardboard disc 11cm across for the Sun, the Earth can be modelled with a ball of Plasticine 1mm across and 12m away. Note that at this scale Pluto is still nearly ½km away!

Setting the homework
Read through the worksheet, making sure the children know what they have to do.

Back at school
Remind the children of their scale model of the Solar System and what they found out in class. **Answers:** 1. Mars, Saturn; 2. a bar drawn that is smaller than the Earth bar but larger than the Mercury bar; 3. about 250million km; 4. Saturn; 5. 2000million km.

Planet	Mercury	Venus	Earth	Mars	Jupiter	Saturn	Uranus	Neptune	Pluto
Size (mm)	½	1	1	½	11	9	4	4	¼
Distance from the sun (m)	5	5	12	18	60	110	220	350	460

My food and drink diary

Foods can be divided into five main groups. Doctors recommend that we should eat a certain number of servings from each group per day.

● Keep a diary of all the food and drink you consume in one day. Encourage your helper to do the same. Fill in the diary below.

● Write down the food groups that each thing you eat belongs to.

Time	Food or drink	Food groups

● Work out the total number of servings from each food group you ate in one day. Ask your helper to do the same, then fill in this table.

Type of food	Recommended servings per day	Total servings per day	
		Me	Helper
Meat and eggs	1		
Dairy products	1		
Cereals	3		
Fruit and vegetables	at least 5		
Sweets	occasional treat only		

● Compare your diet with your helper's.

● How do they compare with the recommended one?

● Why do you think sweets and cakes should only be occasional treats?

Dear Helper,

Your child has been looking at what we need to eat to stay healthy, why we should eat a balanced diet, and what a balanced diet contains. If you can, work through this activity together, keeping a record of everything you eat during the day as well so you can compare diets afterwards. It could be an interesting and memorable experience for you both! For the purposes of recording, one bowl of cereal or a 'helping' on a plate = one serving.

What is digestion?

● Read the following passage and then answer the questions.

Your body needs materials for energy, to be able to grow, and to be able to repair itself. It gets some of these materials from food. The food you eat has to be **broken down** into particles that are small enough to pass into your blood. This is what **digestion** is all about: breaking down complicated materials into smaller particles so they can get into our blood.

Your teeth carry out an important job – they slice and mash your food. While you are chewing, a special chemical, called an **enzyme**, in your saliva (spit) attacks your food, breaking it down into smaller pieces. Your body has many different kinds of enzymes to break down different kinds of foods. If you roll a piece of bread into a ball and chew it for five minutes, the bread should slowly begin to taste sweeter. This is because bread contains lots of a chemical called **starch**, which is like a long chain. The enzyme in your saliva attacks the starch, breaking it down into sugar!

The mashed-up food is squeezed down into your **stomach**. It stays there for a couple of hours, where acidic stomach juices are added, which help to kill bacteria and break down proteins. Muscles in the stomach wall help to mix the food and stomach juices together.

The food then enters a long tube called the **small intestine**, where a liquid called **bile** is squirted in, which helps to break down **fat** into small droplets. The main job of the small intestine is to get the digested particles **into the blood**. It has a very **thin** wall and is surrounded by blood vessels. Once the food particles get into the blood they are taken off to where they are needed for growth and repair, and to give our body energy!

Tough material (such as tomato skins and apple pips) will not be digested. This is called **fibre**, and it passes through the **large intestine**, helping to keep it healthy. The material is still quite wet and sludgy here, so the large intestine absorbs the water before it leaves your body. Your food has travelled almost 9 metres – quite a journey in little more than a day!

1. What does digestion mean?

2. Name two different ways our digestive system breaks up food.

3. In which organ does most digested food enter the blood?

4. Which part of our diet is not digested?

Dear Helper,

This comprehension activity will help your child understand what happens to the food we eat. Most of the ideas have been introduced in lessons at school. Ask your child to read the passage out loud before they complete the questions. As they complete the questions, ask them to explain some of their answers to you – this will help to check their understanding. Do try the bread experiment together if you can!

Gills and lungs 1

● Read this passage.

Although fish live underwater, some of their body systems are very like ours. We breathe the air around us, and fish can use the air dissolved in water to breathe. It may surprise you that liquids can hold gases in them, but think about a bottle of pop. It fizzes because the dissolved gas in it suddenly bubbles out when you open the bottle.

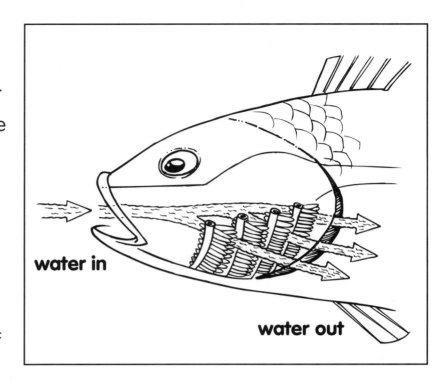

water in

water out

Fish breathe using gills. They draw water in through their mouths. By shutting their mouths they can force the water over the thousands of tiny delicate flaps in their gills. Each flap has a blood vessel close by, and as the water passes over the gills, gases are exchanged between the fish's blood and the water.

When you look at fish you can sometimes see their mouths 'puffing'. When they do this they are drawing water in. Most sharks can't do this, so they have to swim around with their mouths open. If they stop swimming (and so pushing the water through their body) they die – a good reason to keep swimming!

If a fish is taken out of water it cannot breathe. This is because the water supports the delicate gill flaps, and without it they collapse. The fish will puff like mad, but because there is no water, it does them little good.

● Now answer the questions on the 'Gills and lungs 2' worksheet.

Gills and lungs 2

● Look at the diagrams on 'Gills and lungs 1 and 2' and answer the questions.

1. What is the proper name for a human's main breathing tube (A)?

_____

_____

2. What do humans have (B) instead of gills?

_____

_____

3. Which muscle (C) under the lungs pulls down when humans breathe in?

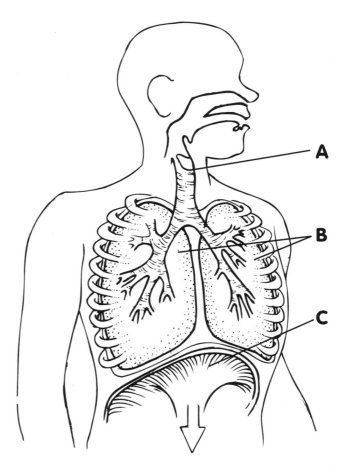

_____

4. Which gas in the air do you think fish take out of the water? (Think about the gas humans breathe.)

_____

5. Which gas do you think fish return to the water?

_____

6. What do you think keeps a fish's blood moving past its gills and round its body? _____

Dear Helper,

Your child has been looking at the human breathing system. After reading this passage, they should be able to recognise similarities between the way we breathe and the way fish breathe. Encourage your child to find out more about the way fish live, by going to the library, looking at a natural history website (www.bbc.co.uk/nature, for example), or by visiting an aquarium.

Does heart rate decrease with age?

● Your task is to try to find out if people's heart rate gets slower as they get older. One way of doing this is to record and compare the heart rate of lots of people of different ages (the more people the better!).

● Once you have decided on your victims (oops, I mean subjects), use this table to record your results, which divides people into different age groups. If your mum, dad and grandparents all claim to be 21, be suspicious!

● Take each person's pulse by holding their wrist like you did at school. Think about how to prepare your subject to find their **resting** heart rate (when they are completely at rest). Count their pulse for one minute, then fill in the table.

Name	Age	Pulse/BPM (beats per minute)

● Try to collect results from a range of age groups. Once you have done this, try to shade in the chart below. Don't worry if you can't fill in a result for every age group.

A chart to show the heart rate of people in different age groups

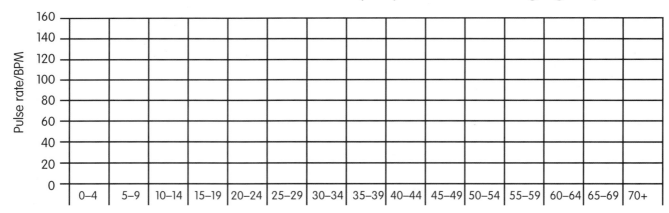

Does heart rate seem to decrease with age? _____

Dear Helper,

This exercise will give your child experience in collecting information for an investigation. Try to involve as many people as possible – friends and neighbours – in the investigation to produce a varied sample of results. Your child might need help with taking and recording the pulse rates, and they might need to practise finding a pulse. Show them how to find the pulse on your wrist by pressing their first three fingers (but not the thumb, as it has a pulse point of its own) until they can feel a beat. Most nine- to ten-year-olds have a resting heart rate of between 55 and 75 beats per minute (bpm).

Changes at puberty

Puberty is a time when young people start to change and become sexually mature.

● Read the statements below. For each one, decide if the change affects only males, only females, or both. Put a tick in the column you think is correct

Change at puberty	Usually happens to		
	Male	**Female**	**Both**
1. Grow more quickly (growth spurt)			
2. Breasts develop			
3. Pubic hair begins to grow			
4. Periods start			
5. Shoulders broaden			
6. Fat is deposited on hips and thighs and the pelvis widens			
7. Extra sebum (oil) is produced by the skin			
8. The voice breaks (deepens)			
9. Facial hair grows			
10. The penis and testicles become larger			
11. An egg is released from the ovaries every 28 days			
12. The body becomes more muscular			
13. The testes begin to produce sperm			

Dear Helper,

At school, your child has been thinking about the physical changes linked to puberty. This is a revision activity; hopefully your child will only need a little prompting about the statements. Talk through any questions they may have while completing the activity. If you are uncertain about the context of this exercise, contact your child's teacher at school.

Smoking

People who smoke are more likely to die of heart disease or diseases of the lungs. The information below has been collected from three countries, and shows how many people out of the 1000 studied in each country died of smoking-related diseases.

● Work out the total and average for each group, then plot a bar chart of the results in the space below.

Number of cigarettes smoked each day	Deaths caused by breathing and heart problems			Total	Average (total÷3)
	Italy	UK	Holland		
0	39	52	29		
10	69	81	39		
20	77	98	50		
30	168	177	105		
40	227	248	155		
50	310	325	268		

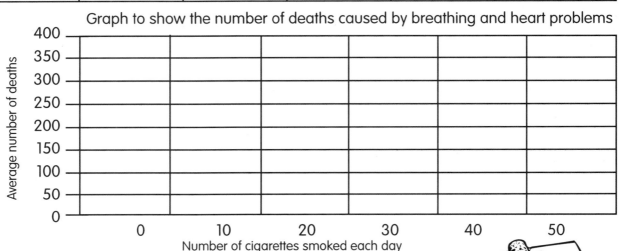

Graph to show the number of deaths caused by breathing and heart problems

Average number of deaths

400
350
300
250
200
150
100
50
0

0   10   20   30   40   50

Number of cigarettes smoked each day

1. What do the results seem to show?

2. It is important to know that the different countries are being compared fairly. What would it be useful to know about the way these results were collected?

3. Heart and respiratory (breathing) diseases are not only caused by smoking. How does your chart help to show this?

Dear Helper,

This activity follows work your child has been doing about how tobacco affects the body. Your child may need help to work out the averages for each group (you can use a calculator if necessary). Explain that they should divide the total by 3 to find the average. Having completed the work, you might like to discuss your and your child's views on smoking.

The life cycle of a bean 1

You will need: your helper's assistance, four broad bean seeds, a 15cm diameter flower pot, enough compost to fill the pot, a space to grow your beans outside (they could grow in garden soil or in a larger pot filled with compost). You'll also need some drawing materials or a camera.

1. Planting the seeds

● Fill your flower pot with compost, to about 5cm from the top.

● Space your beans out on the top of the compost.

Draw or take pictures of your seeds on the compost.

● Cover them with another 2cm of compost.

● Water the compost until it is damp, but not soggy. Leave the pot on a sunny windowsill and give your seeds a little water every other day.

Draw or take pictures when the first shoots begin to appear.

2. Planting the beans outside

● When each bean has grown two strong leaves, it's time to plant them outside.

● Outside, carefully tip your pot of bean plants upside down over some old newspaper, with your hand supporting the soil round the plants.

● Gently separate each bean plant, trying not to damage the roots. Plant them at least 15cm apart in the soil in a sunny corner of the garden, or in compost in a large tub or bucket.

Draw or take pictures of the bean plants in their new position.

The life cycle of a bean 2

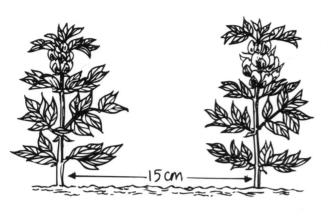

15cm

3. Growing the beans

● Keep the beans well watered and they should grow quite quickly. Watch for the flowers that grow from the stem.

Draw or take pictures of the first flowers that open.

Carefully watch any insect that visits the flowers.

4. Picking the beans

● The flowers will eventually shrivel up and beans will begin to grow in their place. Wait until the bean pods are long and lumpy.

● Pick the pods and carefully open them up: inside should be some large bean seeds. Save some of the beans and keep them in a dry place so more beans can be grown next year. Ask an adult to cook some of the beans so you can eat them.

Draw or take pictures of a pod containing seeds.

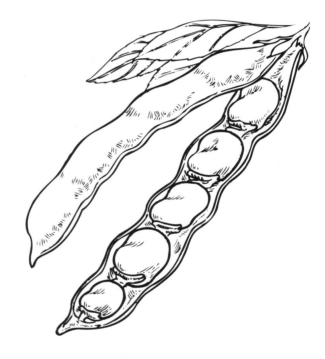

Dear Helper,

This practical project will help your child experience and understand the life cycle of a flowering plant. Your child will need help with both the planting and transplanting of the seeds. Encourage them to keep a pictorial record of the different stages of the plant's growth. If you're successful, enjoy eating the beans together and planting next year's crop!

Tree seed detective

Mr Seed's seed company collects tree seeds and sells them in small packets. The packets are all the same size, so they can fit lots of small seeds into each packet but only a few large seeds. Of course, if the seeds were not collected they would be dispersed naturally. They could be wind-blown, bird-sown, or heavy-hoarded seeds.

● Look at this list of trees. Look carefully at the number of seeds the company could fit in a packet.

● From each description, decide if the seeds would be **wind-blown**, **bird-sown** or **heavy-hoarded** in nature. The first one has been done.

Tree	Description of the seed or fruit	Number in each packet	How is the seed dispersed?
Maple	Green and red winged	80	*wind-blown*
Walnut	A plum-sized, wrinkled nut	5	
Birch	Tiny, like confetti	400	
Hawthorn	A small nut inside a fleshy shell	25	
Sycamore	A double-winged key shape	80	
Chestnut	Large, shiny, brown nuts	4	
Holly	Red berries around a small nut	20	
Dogwood	Encased in a black berry	30	
Hazel	Brown shell around a white core	5	
Ash	Clusters of winged keys	80	
Beech	Smooth, shiny brown nuts	5	
Bird cherry	Purple berry around a tiny nut	20	
Alder	Small, papery-winged	100	

● What do you notice about your answers for the packets containing the greatest numbers of seeds?

Dear Helper,

In order for new plants to grow, seeds need to be dispersed from the parent plant. This can happen in many ways, which your child has been looking at in class. Go for a walk in the park or the countryside and look for trees to see how the seeds might be dispersed naturally before completing this sheet. Remind your child that berries can contain poisonous seeds, and should not be eaten.

Insects and flowers

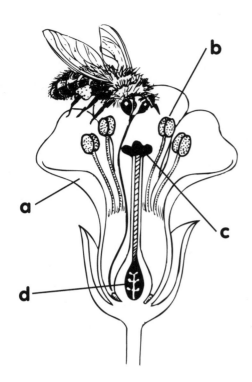

a
b
c
d

● Here is a diagram of a flower. Write the name of the part of the flower by each question.

a This part of the flower is designed to attract insects. _____

b Insects will find pollen on this part of the flower. _____

c Insects transfer pollen to this part of the flower to enable seeds to grow.

_____

d Seeds grow in this part after an insect has pollinated the flower. _____

● Here are eight sentences that describe what happens when a flower is pollinated, but they've been mixed up. Write the letters in the correct order. We've given you the first and last answers.

a The insect searches inside the flower for its food of pollen and nectar.

b Some pollen sticks to the body of the insect.

c Fertile seeds will now grow in the ovary of this flower.

d Petals attract an insect to a flower.

e The insect rubs against the pollen on the stamen.

f The insect flies off to another flower.

g Some of the pollen on its body rubs onto the stigma of this flower.

d
c

● Write a sentence to explain why gardeners might want to encourage some insects to their gardens.

Dear Helper,

We have been talking in school about how insects (such as bees) transfer pollen from one plant to another to pollinate the flowers and produce seeds. Help your child by looking carefully at a real flower (or a picture of a flower) and asking them to identify the parts labelled in the diagram at the top of the page.

Frogs in the garden 1

● A gardener wrote a letter to the editor of a gardening magazine. It said:

> Lots of the plants that I have tried to grow in my garden have been eaten by slugs. Can you help me get rid of these pests using natural methods? I don't want to use poisons to kill the slugs as these might harm other creatures.

● The editor of the magazine wrote this reply:

> I am sorry to hear that you have lots of slugs in your garden. They can damage many young plants. However, I am really pleased you want to get rid of these pests without using poisons like slug pellets.
>
> One really useful creature to encourage into your garden is the frog. You could build a small pond in your garden and fill it with water and plants. In spring, ask another gardener for some frogspawn and put it in your pond. After a few weeks, small tadpoles will develop. The tadpoles will feed on other creatures and plants that have made their home in your pond. Eventually, the tadpoles will develop into small frogs, who will leave the pond and live in your garden. They'll spend most of the day hiding in dark, damp places under plants, but at night they'll come out and eat the slugs!
>
> After two or three years, the frogs will develop into adults. They'll return to your pond and mate. If you're lucky your pond will be full of new frogspawn.
>
> Once you have a population of frogs living in your garden, slugs will be less of a problem!

■SCHOLASTIC

Name:

Frogs in the garden 2

● Answer these questions.

1. Why might gardeners want frogs living in their garden?

_____

_____

2. Here are some sentences about the life cycle of a frog. Write them in the right order, starting with 'Male and female frogs mate'.

a Male and female frogs mate.   **b** Tadpoles grow into young frogs.

c Young frogs leave the pond.   **d** Frogspawn develops into tadpoles.

e Young frogs slowly become adult.   **f** Frogspawn floats in the pond.

_____

_____

_____

_____

_____

3. Frogs and tadpoles are eaten by many creatures. Use reference books to make a list of some of the enemies of the frog.

Dear Helper,

We have been talking about plant and animal life cycles at school; this activity introduces the life cycle of the frog. Your child will need access to reference material (from the library or the Internet) to complete the last question. Help your child with the research if necessary, reading the text through together if they find the language difficult.

PHOTOCOPIABLE

Where has the song thrush gone?

Lots of people know that creatures like whales and tigers are in danger of extinction, but did you know that in parts of our country some species may become extinct too? One bird, the song thrush, is now quite rare.

When song thrush chicks hatch out of their egg they are naked and blind. They are quite defenceless, and many get eaten by other creatures.

Many farmers have dug up hedges between fields so they have more space for growing things. Gardeners and farmers have been spreading pesticides that kill the snails that eat their crops!

● Find out the answers to these questions about the song thrush using reference books or the Internet.

● Write your answers on the back of this sheet.

1. About how long is an adult song thrush?

2. What colour are the feathers on its back and the spots and streaks on its white chest?

3. Make a list of some of the different kinds of food the song thrush eats.

4. How does the song thrush manage to eat snails that live in hard shells?

5. In what kind of places do adult song thrushes make their nests?

6. Make a list of creatures that prey on song thrush chicks.

7. Why are digging up hedges and poisoning snails problems for the song thrush?

Dear Helper,

It is common to think of dinosaurs and dodos as 'extinct', and animals such as tigers as 'rare'. However, many once common local creatures are also at risk. In Britain, the song thrush is an endangered species. Help your child to research the answers to these questions using reference books or the Internet, or at the local library. See if your child can use the contents and index pages of books to locate information.

Endangered species in our skies

● Here is some information about common British birds.
There are far fewer of each of these species than 25 years ago.

	The number of each species living now compared to 25 years ago	How likely you are to see one of these birds in summer in Britain
Blackbird	About 7 now for every 10 then	88%
House sparrow	About 4 now for every 10 then	60%
Starling	About 5 now for every 10 then	71%
Song thrush	About 5 now for every 10 then	65%
Swallow	About 8 now for every 10 then	70%

Data from the British Trust for Ornithology (www.bto.org)

1. For each species, look at the table and then cross out the number of birds that would not be alive today compared to 25 years ago.

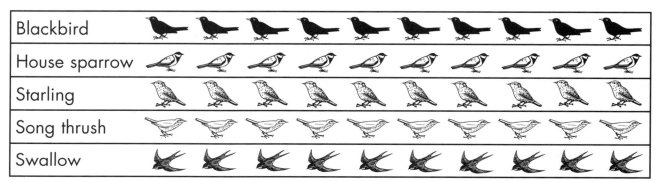

Blackbird	
House sparrow	
Starling	
Song thrush	
Swallow	

2. a Which of these five species of bird is causing bird-watchers the most concern? _____

b Two other species are a serious worry to bird watchers. What are they?

_____

3. The last column in the table tells bird-watchers how likely they are to see each species if they go bird-watching in summer.

a Which of the birds is an ornithologist most likely to see? _____

b Which species are they least likely to see? _____

Dear Helper,

Your child has been thinking about endangered species in science lessons. These figures have been adapted from studies of endangered British birds made by the British Trust for Ornithology. This activity should help your child practise their data-handling skills, particularly understanding what percentages mean. It's important for children to see how maths is used in other subjects such as science. Talk through the calculations with your child if you need to. Watch out for percentage figures used on TV or in newspapers.

The life cycle of a border collie

Border collies are a breed of dogs that have been specially bred to help farmers. They are hard-working animals that seem to enjoy rounding up sheep. They often work in harsh environments.

● Here are the seven stages in the life cycle of a border collie.

a The puppies feed on their mother's milk.

b Over 63 days each embryo develops into a puppy.

c The puppies are weaned and begin to feed independently.

d Inside the female, eggs are fertilised and become embryos.

e A litter of as many as seven puppies is born.

f Adult male and female border collies mate.

g It takes a year for a puppy to mature into an adult dog.

1. Complete the life cycle diagram for a border collie. Put the letter showing each stage in the correct box.

2. Look again at the sentences that describe the life cycle of a border collie. Find these words and underline them in the sentence, then write what you think it means in the space below:

weaned _____

embryo _____

mature _____

3. Why do you think border collies, and other dogs, produce so many puppies?

Dear Helper,

This activity aims to improve your child's understanding of the life cycle of mammals. Help your child use a dictionary to find the meanings of the words in question 2 if they get stuck. To help your child think about question 3, ask them what difficulties young animals might have in surviving, especially if they weren't looked after by humans.

Name:

The water cycle in the kitchen

● Look carefully at the picture and answer these questions.

1. How does clean fresh water come into the kitchen?

_____

_____

2. Can you see three places in the kitchen where liquid water is evaporating into water vapour? _____

3. There is so much water vapour in the air in the kitchen that some is condensing. Where is this condensation taking place?

_____

4. How is water vapour being removed from the kitchen?

_____

5. How is dirty liquid water removed from the kitchen?

_____

6. Water can either be a **liquid**, a **gas** (water vapour) or a **solid**. Where might you find water as a solid in this kitchen? _____

Dear Helper,

Water travels through the environment in a process called the *water cycle*. This homework revises elements of this – *evaporation* and *condensation* – and shows where we can see the water cycle in the home. Ask your child to explain these two terms to you, and see if you can spot these processes at home. Add local detail to your child's understanding by helping them to find out where your water supply comes from and where the dirty water is treated.

Name:

Getting rid of water

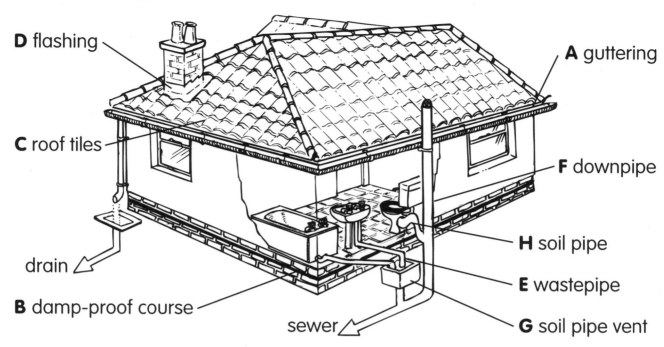

D flashing

C roof tiles

drain

B damp-proof course

sewer

A guttering

F downpipe

H soil pipe

E wastepipe

G soil pipe vent

● Look at this picture of a house. Different parts that help collect and carry water away from it have been labelled.

● Match each of the sentences below to the correct part of the house.

1. These are overlapping flat surfaces that prevent rain from getting inside the house.

2. This takes dirty water away from baths and sinks.

3. This carries waste and dirty water away from toilets and into sewers.

4. These collect water that has run off the roof.

5. A flexible, rust-proof metal that keeps rain out of the gullies around chimneys and windows.

6. This carries rainwater from the gutters to the drains.

7. This stops water and dampness in the ground soaking up into the walls.

8. This allows smelly unpleasant gases to escape into the atmosphere.

● Find out more about the damp-proof course in a house. Find at least three different things we use to stop water soaking up into the walls of houses.

Dear Helper,

There are many familiar things inside and outside our homes that keep the building dry (roof tiles and guttering, for example). This activity will encourage your child to think about and understand the purpose of these objects. Go out with your child and try to spot the features in the picture, and discuss their purpose to complete the sheet. Help your child to find information on damp-proof courses from DIY books or from the local library.

Name:

Saving water saves money 1

● Read this leaflet. A water company sent it to the headteacher of Big Hill primary school.

Dear Headteacher,

We supply the water to your school. We measure how much water your school uses with a water meter. Every drop of water you use has to be taken from the natural environment, and this has an effect on other animals and plants. The water you use is expensive, so it makes good sense to use it carefully.

Here are some ways you can use less water at school:

● Teach your children that clean, fresh water is precious and shouldn't be wasted.

● Persuade children to turn off all taps when water is no longer needed.

● Make sure dripping taps are mended.

● Insulate and protect water pipes so they don't burst in cold weather.

● Fit spray taps and toilets that use less water.

● Collect and use rainwater for school ponds and gardens.

Here is a good way to make sure that there are no leaks in the pipes in your school:

1. When the children have gone home, switch off all the taps and any other things that use water.

2. Read the number on the water meter.

3. Several hours later read the number on the water meter again.

If there are any leaks in your pipes, the number on the meter will change.

Name:

Saving water saves money 2

● Use the information in the letter to answer these questions.

1. Why it is important that we do not waste clean, fresh water?

_____

_____

2. The caretaker makes sure the school building is safe and in good condition. List two things that the leaflet asks the caretaker to do.

_____

_____

3. The headteacher has to make sure that the school's money is spent wisely. Suggest two things the headteacher might decide to do to save money on the school water bill.

_____

_____

4. The school caretaker follows the instructions on the leaflet and checks for leaks in the school water system. The two meter readings are shown below – one at 6 o'clock and one at 9 o'clock. What does this tell the caretaker?

_____

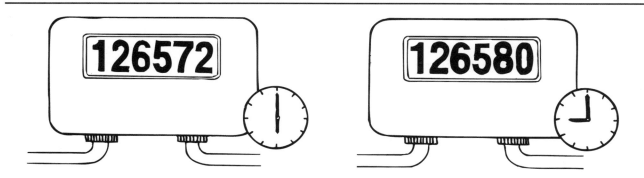

Dear Helper,

Water is a precious resource, and producing clean, fresh drinking water is expensive, using fossil fuels that can damage the environment. This activity will help your child understand some of the practical issues involved in water conservation. Read through the worksheet together, and discuss your child's answers with them. Challenge them to suggest ways of conserving water at home.

A diary of a mini-pond

You will need: a small container (such as an empty margarine tub), water, writing paper, a magnifying lens.

● A great variety of creatures need to live in clean, unpolluted water. Follow these instructions to make a tiny unpolluted pond near your home, then keep a diary of the creatures that use your pond.

How to make a mini-pond

Make your pond out of a small container, such as an empty margarine tub. With your helper, find a sensible, safe place outside your home to keep your mini-pond. It needs to be in a place where it won't cause an accident, but can be useful for wildlife. Make sure you can see your mini-pond from inside the house. Fill the container with water (rainwater is best). You can add things like stones, soil or dead leaves to encourage visitors to your pond.

What lives in your pond?

When you have made your pond, try to watch it carefully at the same time each week. For example, every Sunday morning you could watch your pond from a window for ten minutes. Try to identify any bird or creature that visits your pond for a drink. Keep a record of what you saw in your diary.

Then go outside and look carefully at your mini-pond. There may be some tiny

> **MY MINI-POND DIARY**
>
> Date   What I saw in my mini-pond
> Sunday   I put my pond in the garden near my mum's
> 20th   vegetable patch. I filled it with rainwater
> April   from the barrel behind the shed. I watched
>   for ten minutes from the kitchen. A blackbird
>   went near it but didn't touch the water.
>
> Sunday   I watched my pond from the kitchen window.
> 27th   A sparrow went up to my pond and had a
> April   quick drink. I went outside and looked in the
>   pond. I don't think anything is living there.

creatures living in it (this is when a magnifying lens is really useful). Try and draw some of the creatures you have seen in your diary.

Of course, the water in your pond may evaporate. If this happens, add more water to your pond. Keep a note in your diary each time you fill your pond up again.

Dear Helper,

This is a safe activity that gives your child an opportunity to take an interest in local wildlife. It also allows your child to practise the scientific skill of making regular and detailed observation. The mini-pond is more likely to be successful if you share the task with your child and together try to identify the creatures that use the mini-pond. Discuss where might be the best place for the pond – a sheltered place close to plants usually attracts insects and birds well.

Reducing pollution

We all use lots of electricity – it's wonderful stuff! As well as lighting and heating our homes and schools, it makes things like televisions, computers and CD players work. However, most of the electricity we use is made in power stations that burn coal, gas or oil to make the electricity. As these fuels are burned, they produce air pollution which can cause **acid rain**, and may cause **global warming**.

● Complete these sentences, which explain the chain of events.

1. Most of the electricity that we use is made in _____ _____.

2. Most of these burn coal, gas or oil which _____ the air.

3. This causes _____ _____, which can destroy trees and kill fish.

4. The pollution may cause _____ through global warming.

5. Look around your bedroom and complete the table.

	Tick if you have one of these electrical appliances	Do you always switch it off when you are not using it? Yes or No
Ceiling light		
Bedside light		
Other light		
Television		
Tape/ CD player		
Radio		
Computer		
Other		

● Look at your answers. Is there anything you could do to use less electricity and help reduce air pollution?

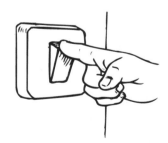

Dear Helper,

This activity will help your child think about how everyday activities may inadvertently cause damage to the environment. The burning of fossil fuels to generate electricity has caused acid rain and may cause global warming, leading to flooding of low-lying countries such as Bangladesh. Talk with your child about ways in which you could use less electricity at home.

Name:

Has the cod had its chips?

You will need: Some centimetre squared paper, a pencil, ruler and coloured pens or pencils.

● Do you eat fish and chips?

● If you do, the fish you eat is likely to be **cod**. Lots of cod are caught in the North Sea by fishermen from Britain, Denmark, Holland, Belgium, France, Germany and Spain. Each year the weight of cod caught by all these fishermen is measured.

● This table shows the weight of fish caught every other year:

Year	Weight of fish caught/tonnes
1981	340 000
1983	260 000
1985	220 000
1987	200 000
1989	140 000
1991	100 000
1993	120 000
1995	160 000
1997	120 000
1999	100 000

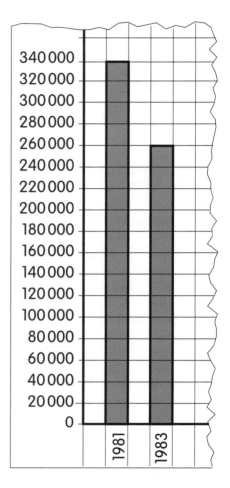

1. Make a bar graph of the amount of cod caught each year. Use one square for every 20 000 tonnes of cod caught. The graph may help you.

2. Look at the shape of your bar graph. What does it tell you about the amount of fish caught each year?

3. Why do you think that fewer cod are being caught now than 20 years ago?

Dear Helper,

This activity draws on your child's knowledge of making graphs to look at the decline of cod fish stocks in the North Sea. Help your child if necessary by reading the large numbers with them: *Three hundred and forty thousand tonnes* to *One hundred thousand tonnes*. Discuss possible reasons for the lower tonnage caught in more recent years: think about all the countries sending fishermen to fish the North Sea.

Name: _____

Gas to liquid

● You can use the words below once, more than once or not at all:

gases	shape	squash	squashed	liquid	solid	runny	air

1. A brick is a _____. It cannot be _____, and keeps its _____. Water is a _____. It is _____ and difficult to _____. It will take up the _____ of the container it is in. We can fill balloons with air or helium. These are both _____.

2. a Draw what will happen when you let the air out of balloon **X**.

b Why will this happen?

c What would you do to make it a fair test?

d Draw what you would see if you filled both balloons with air, but balloon **X** contained more air than balloon **Y**.

e Draw what would happen if balloon **Y** was filled with air, and balloon **X** was filled with the same amount of helium.

Dear Helper,

In class, your child has been looking at describing and defining the three 'states' of matter – solids, liquids and gases. This homework is revision of work done in class. Your child has already looked at the balloon experiment at school, so they should be able to answer the questions without setting the experiment up again. Challenge them to explain the difference between a solid, a liquid and a gas to you.

PHOTOCOPIABLE

Name:

Air

We are surrounded by air, which is a mixture of different gases. It is invisible, but we couldn't survive without it.

● Find out as much as you can about the air that surrounds us. Make some notes, then write a report.

● The answers to these questions should help you write your report:

What is the air made up of?

Who discovered the different gases in the air? When?

Why are the different gases in air important to life on Earth?

If you took a car to the Moon would it work? Why?

Where will you find the information that you need?

Dear Helper,

During our science lessons, we have been learning about air. This activity asks your child to find out more about the air we breathe in order to write a report. The reference section of your local library will have books which will be a good source of information. The Internet is also a useful research tool, but if you do have access to a computer, make sure your child is aware that simply downloading information they don't understand is not beneficial – taking notes is the key to good report writing.

Name:

Solid, liquid or gas?

Find an adult to help you with this homework. Never mess with any gas! If you find a liquid and you are not sure what it is, ask an adult – some liquids are poisonous and may burn you.

● Look around your home and find five (or more) materials. You must find at least one example of a gas and at least two examples of a liquid.

● Fill in the table below.

Material					
Does it have a fixed shape?					
Do you think that it can be squashed?					
Does it spread out into the air when it is released?					
Is it runny?					
Is it a solid, a liquid or a gas?					

● Fill in the two boxes to show the arrangement of particles in a liquid and in a gas.

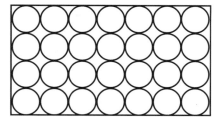

A solid   A liquid   A gas

Dear Helper,

This activity aims to reinforce the work your child has done in class on the properties of solids, liquids and gases. Help your child get started by looking around your home together, asking them to point out some materials that are solids, some that are liquids and some that are gases. Ask them to explain how they know which is which as they fill in the table.

Steamy

1. If you want to turn a bowl of water into steam, what must you do to the water? _____

2. a Is water a solid, liquid or a gas ? _____

   b Is steam a solid, liquid or gas ? _____

3. If you wanted to change steam back into water, what would you have to do to the steam? _____

4. What is the name of the process that occurs when a gas, such as steam, changes back into a liquid, such as water? _____

5. Look at the diagrams below. Which arrow shows condensation? _____

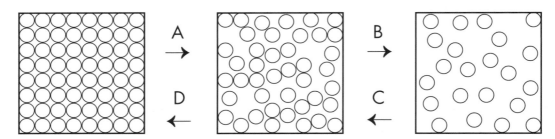

6. Which of the following are examples of condensation? Tick the boxes that you think are examples of condensation.

☐ **a** 'Steamed up' kitchen windows

☐ **b** Water turning into ice

☐ **c** Water droplets on the outside of a glass of cola with ice

☐ **d** Butter turning to a liquid in a hot pan

☐ **e** A chocolate bar going soft and runny

Dear Helper,

This activity revises work your child has been doing in class about how materials can change from a liquid to a gas. When your child gets to question 5, challenge them to explain each of the diagrams and to tell you what the arrows A–D represent. (The three pictures represent a solid, a liquid and a gas respectively; arrow A represents melting, B is evaporating, C is condensing and D is freezing.)

Liquid to gas

- Mrs Brown was making Sunday lunch. She wanted to boil some potatoes, so she boiled a pan of water before cooking the potatoes.

- She decided to carry out an experiment and measured the temperature of the water every minute until the water started to boil.

- Here are her results.

Time/min	0	1	2	3	4	5	6	7	8	9	10	11	12
Temperature of the water/°C	21	26	36	49	61	73	81	90	98	100	100	100	100

- Draw a line graph of Mrs Brown's results.

Dear Helper,

We have been looking at what happens to the temperature of water as it is heated to boiling point in class. This activity reinforces the work your child has done in class, and allows them to practise plotting a graph from a set of results. It is very important that your child does not carry out this activity for themselves – take this opportunity to reinforce the dangers of boiling water.

SCHOLASTIC 67

UNIT 4 MATERIALS GASES, SOLIDS AND LIQUIDS

PHOTOCOPIABLE

Name:

What a state!

● Complete these sentences using words from the box below. Each word may be used once, more than once, or not at all.

gases	liquids	boiling	pieces	big	small
melting	solids	gas	liquid	solid	

The three states of matter are _____, _____ and _____. _____ and _____ are harder to compress than _____. _____ and _____ have a fixed volume, but _____ do not. _____ and _____ can flow, but _____ have a fixed shape. The temperature at which a solid turns into a liquid is the _____ point. The temperature at which a liquid turns to gas is the _____ point.

● This plastic bottle has cherryade in it. Write in the boxes to show the parts that are:

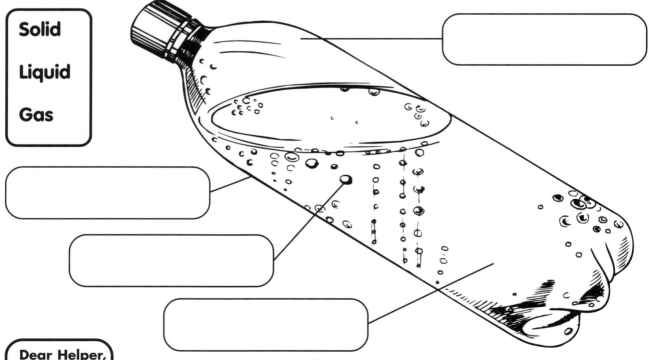

Solid

Liquid

Gas

Dear Helper,

When you heat or cool a material, you can cause it to melt, evaporate, freeze or condense (think about what happens when you heat or cool water to make steam or ice). When your child has completed these sentences, you could check that they have spelled the words correctly, and help them with words they have had difficulty with by asking them to read the word, cover it up, write it out and then check it.

Solid to liquid

- Fill in the missing words in the sentence below.
  Use each word or phrase in the box below only once.

| melting point | freezing point | melts | liquid |

If you heat a solid, you can turn it into a _____. We say that the

solid _____. The temperature at which this happens is called the

_____ _____. This temperature can also be called

the _____ _____.

- Look at the table below.

Substance	Melting point/°C
Aluminium	660
Ice	0
Iron	1535
Copper	1083
Polythene	110

1. a Which material has the
lowest melting point?

_____

b Which material has the
highest melting point?

_____

2. Room temperature is about 20°C. Which of these materials may be a
liquid at room temperature? _____

3. When heated from room temperature, which melts first: aluminium or
copper? _____

4. What happens if you keep heating a liquid? _____

- On the back of this sheet, draw a bar chart of
  the melting point values in the table.

Dear Helper,

Your child has carried out an investigation to find out how long ice can keep a drink cold. The freezing
and melting point of water is 0°C, and ice in a drink at room temperature will slowly melt until the liquid
reaches the same temperature as the liquid around it. This activity continues the theme of melting and
freezing points and, in the second part, emphasises the fact that different materials have different melting
points. Challenge your child to list four things which they have seen melt (chocolate or butter, for example),
but it is not wise to let them try melting things for themselves.

PHOTOCOPIABLE

Name:

Going on and on and on...

You will need: a ruler and a pen or pencil.

In the shops you can buy lots of different types of batteries: different sizes and different makes. Batteries have lots of uses, but they will last for different amounts of time when used in different electrical items. The makers of batteries are always keen to advertise how long they will last.

Make of battery	Hours of use in a torch
Econobatt	4
Everuse	8
Everuse Gold	12
Forevercell	16
Megajuice	10

● This table shows how long five different makes of battery will last when used in a torch.

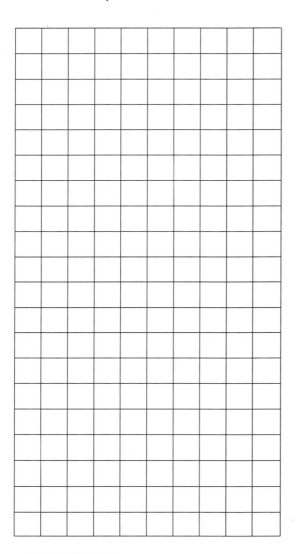

● Plot the information from the table on a bar chart using the graph paper, then answer the questions.

1. Which battery lasts the longest in this torch? _____

2. How many 'Econobatt' batteries would you use up for one 'Forvevercell' battery?

_____

3. For £2.99 you can buy a pack of six 'Everuse' batteries or a four pack of 'Megajuice' batteries. Which set of batteries would allow you to use your torch the longest? _____

Dear Helper,

Transferring information from a table to a chart is an important skill for your child to master in maths and for science investigations. This exercise uses a scientific example to practise this skill. Your child has been learning about batteries and mains electricity in class. The questions require your child to investigate the data further.

Name:

Locked out

● This diagram shows an overhead view of a simple door lock that is opened using an electromagnet. When the switch is pressed down, the steel bolt is pulled back by the magnet so that the door can be pushed open.

● Look at the diagram carefully and then try to answer the questions.

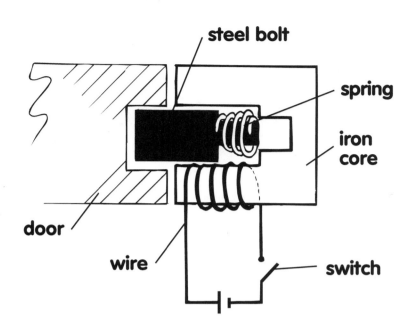

steel bolt

spring

iron core

door

wire

switch

1. Pressing a switch completes what? _____.

2. What goes around the wire when the switch is pressed down?

_____

3. Why does the steel bolt move back when the switch is pressed?

_____

4. If the steel bolt was replaced with an aluminium bolt, the lock would not work. Why? _____

5. What would happen if the battery went flat? _____

6. What could you add to the circuit of this locking system to tell the person pressing the switch that the battery was going flat?

_____

Dear Helper,

This activity reinforces our recent work in class on electromagnets, and should also help your child to revise electricity and magnetism work done in an earlier topic at school. An electromagnet becomes magnetic only when electricity is passed through it, hence it can be 'switched on' to attract the steel bolt in the door lock. Challenge your child to tell you about places in your home where electromagnets might be found. The answer to question 6 requires some type of indicator or warning device to be fitted to the lock that is triggered when the battery begins to run down.

PHOTOCOPIABLE

Spinning around

Electricity can be used to make light using bulbs, to make sound using speakers or to make movement using motors.

● With your helper, look around your home and try to find as many things as you can that use a motor. Make a list of all the motors you find in each room.

Kitchen
Lounge/dining room
Bathroom
Bedrooms
Garage/garden shed

● As you make your list, try to decide which motor might be the biggest by looking at how much work they have to do (perhaps because they are moving heavy loads or spinning very fast).

Dear Helper,

It's important for your child to realise how many different uses there are for electricity in real life – when you're out, see how many different uses your child can spot. At school, we have been looking at how motors work. This homework gives your child the opportunity to combine two important aspects of science learning: scientific knowledge (or *theory*) and real-life uses (or *application*).

Name:

Simon's new crane

- Simon got a toy crane for his birthday.

- Unfortunately, the instructions of how to make it work were missing, and the only information he has to decide how the crane might work is this diagram of the electric circuit it uses.

- Use the information on the circuit diagram to write a set of instructions for Simon on the back of this sheet.

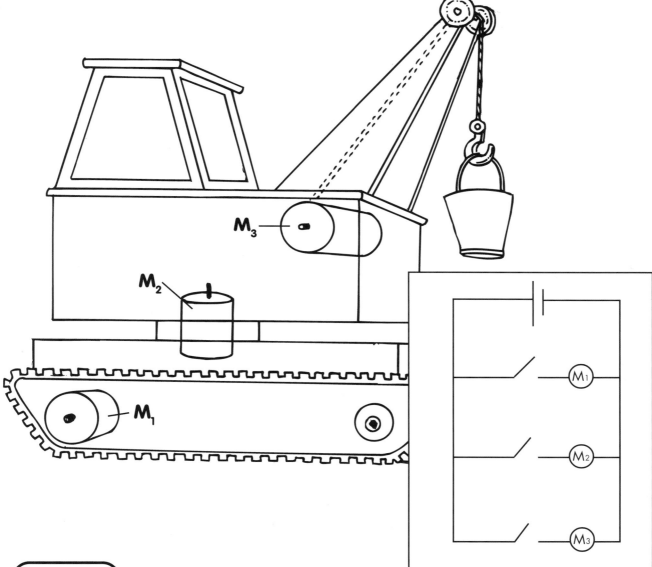

Dear Helper,

We can use a circuit diagram to explain how an electrical circuit works using a standard set of symbols. Interpreting a circuit diagram and describing what it does from a picture is a skill commonly tested in national tests. Here, it is rather more tricky in a 'real-life' situation. If your child gets stuck, prompt them by asking what the 'M's mean (they are motors), and to show you where the switches are on the circuit and what they may be for.

We have the power

There are many power stations in the country, and many more homes, offices and factories that use electricity. We need different amounts of electricity at different times of the day. What would happen if the power station nearest your house wasn't working? Would you have no electricity? Try to find out how these problems are sorted out.

● Use the space below to make notes about what you find out. Here are some key words to get you started on your research.

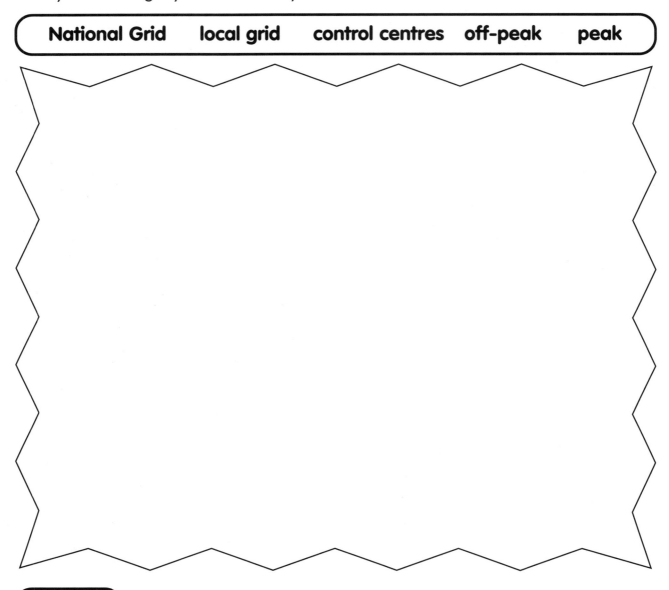

| National Grid | local grid | control centres | off-peak | peak |

Dear Helper,

This activity requires your child to do some research. They might want to use the local library, encyclopedias, CD-ROMS or the Internet – even electricity bills may be useful, as they often come with information printed on or with additional leaflets about supply. Try to encourage your child to use only information that they understand; it is all too easy to find pages of irrelevant and unintelligible information and simply copy them, but this won't help them understand what they find out. If your child is keen and finds out lots of information, encourage them to write a short report using the information they have collected.

Name: _____

Emission control

● Use the words in the box to fill in the blanks in the passage about how electricity is made. The words in the box can be used once, more than once or not at all.

pollute	burning	air	carbon dioxide	tides
Sun	rivers	waves	generator	fuel

In the UK, most electricity is produced by _____ fossil fuels, namely coal, oil and natural gas. This heats up water to make steam, which turns a turbine and turns a _____. The turbine looks like a windmill.

When the _____ is burned it produces a gas called _____ _____. This causes pollution in the environment. To look after our environment, we should use less electricity so that less fuel has to be burned, and we could use alternative forms of electricity production.

Solar power uses energy from the _____, which is converted into electricity using special 'solar panels'. We can use wind turbines to collect energy from the movement of _____, and in mountain areas we can use the energy of the water in _____ to turn turbines. We can use the energy of the _____ and _____ in the sea to make electricity too.

These types of power station do not produce as much electricity as fossil fuel power stations, but they do not _____ the environment. Unfortunately, the best locations for these friendly power stations are usually the prettiest parts of the country, for example mountains and rivers.

Dear Helper,

This exercise combines interacting with a text, an important literacy skill, with using scientific vocabulary. It also looks at some important environmental issues that are considered in other science topics and PSHE (personal, social and health education) and citizenship. Discuss your ideas and opinions on the environment and alternative energy sources with your child, and try to look at both sides of the argument.

PHOTOCOPIABLE

Name:

Mind the gap

There are many different reasons why bridges are built: to cross small streams or big rivers, or to cross roads or railways. There are many different shapes and styles of bridge, because of their age, size or the building materials used.

● Try to find out about some different types of bridge, and draw a labelled picture in the space below.

● Here are some words to help your research:

| beam | arch | suspension bridge | cantilever bridge |

Dear Helper,

The information needed to complete this activity can be found in reference books, on CD-ROMs or through the Internet. Try to discourage your child from printing out or copying pages of information where only a small part is relevant. Help them to take notes and develop summarising skills before drawing their annotated picture. If your child is interested and finds a lot of information, suggest that they make a small project file using the information they have found.

A day out at Adventure Island

Adventure Island is a fun place to go. Here is the advertising leaflet that tells you all about the things to do. There's a map of the island to show you where to find them.

ADVENTURE ISLAND

Come and spend time relaxing and having fun on Adventure Island. Explore the beach like Robinson Crusoe! Track deer and squirrels and spot the birds in our woodlands like Robin Hood! Or follow the stream and tickle the trout like Huckleberry Finn.

A Crusoe's Beach – roam the sands and explore the rock pools. Boat trips around the island daily, every 30 minutes.

B Adventure Forest – Visit the hides to see how the wood is home to deer and many birds. How many kinds of tree can you spot?

C Tree Town – Are you brave enough to climb up to the viewing platform high in the treetops? There are ropes and tyre swings there for the Tarzan-types among you!

D Deep River – Cross by the bridge and paddle at the bend in the stream. Beware: the water, weed and algae here can make the stones and bank path slippy underfoot. Ask a Ranger to point out the fish and other wildlife.

Adventure Island: where every visit is an adventure!

● Look at the descriptions. What sort of materials do you think you would find in each place that could affect friction? How would these materials change the force of friction?

Dear Helper,

Relating classroom science to real life is very important. All around us there are materials to reduce friction (such as grease on axles) and materials that increase friction (such as rubber car tyres). This activity aims to show that there are similar materials in the natural world. Encourage your child to think of at least one material that might change the force of friction in each place described on the leaflet. Look for rough, smooth and slippery materials next time you are out with your child; you might like to try using crayons to make a bark rubbing of an interestingly patterned, rough tree trunk.

Loosen up

You will need: a pen or pencil, something to lean this sheet on and an adult or two to tell you their stories.

Lubricants reduce the force of friction. They make things slippery. This can be very helpful if something has got stuck!

● Try to find out when and where lubricants have been used in your home. You may be told some funny stories, but try not to laugh too much – you might be the one who gets stuck next time! Try to find out why things got stuck in the first place and which lubricant was used to loosen them.

● Write down your stories in the space below. You could write notes or make a table – remember that you don't have to write down all of every story.

Dear Helper,

Every family has a story of the day when Auntie was a little girl and got her hand stuck in the sweetie jar, or something similar. Help your child to find family or friends that have some good real-life 'disaster' stories, hopefully with a happy ending! Concentrate on how the problem was solved by using something slippery such as soap or washing-up liquid. Ask your child to explain how the lubricant will have helped.

Ball-bearing clocks

● This diagram shows a ball-bearing clock. Look at it carefully and then answer the questions.

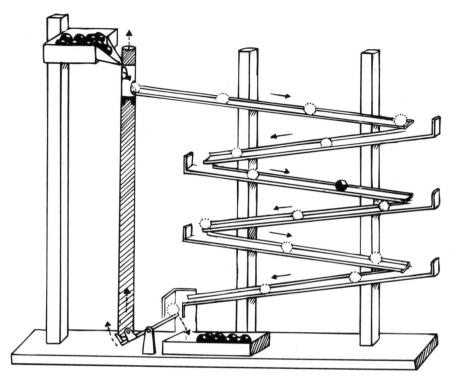

1. What force is making the ball-bearing fall? _____

2. What force will slow the ball-bearing down on the way along the runners? _____

3. When one ball-bearing gets to the bottom, how does the clock start the next ball-bearing in motion? _____

4. It takes a ball-bearing 20 seconds to travel from the top of the clock to the bottom. How many ball bearings will fall in one minute?

_____

5. The tray at the top of the top of the clock can hold a maximum of 15 ball-bearings. What is the longest time the clock can work for without being reloaded? _____

Dear Helper,

Your child has been making timers in class. This activity looks at another kind of timer, and reinforces the work your child has done recently in school on forces. It requires them to look at the clock diagram in detail to consider the forces involved and how they act on the ball bearings. Look at the diagram together, and encourage your child to think about how the forces of gravity and friction are affecting the ball-bearings as they move down the clock.

SCHOLASTIC

Sail racing

You will need: a pencil and/or pen, a ruler and a piece of graph paper for this activity. Your teacher will tell you whether calculators are allowed!

● Kelly and her family were racing boats across a large paddling pool in their back garden.

● Each boat had a different-sized sail, and Kelly timed how long it took each boat to cross from one side of the pool to the other.

● The table below shows the sizes of sails used on model boats that are raced across a large paddling pool and how long it took each boat to get from one side to the other.

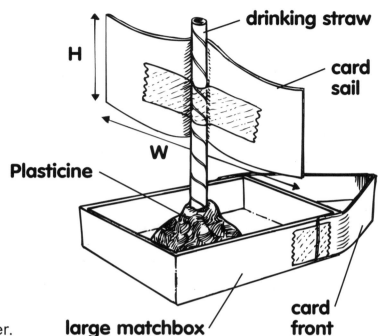

● Work out the area of each sail, then use this information to plot a line graph showing how the area of the sail affects the time it takes the boat to cross the paddling pool.

Width of sail (cm)	Height of sail (cm)	Area of sail (cm²)	Time of journey (s)
5	5		60
5	10		45
10	10		30
10	20		20
20	20		15

Dear Helper,

This activity reinforces two skills: in numeracy, using a formula to calculate the area of a rectangle, and in data handling the transferring of information from a table to a line graph. Make sure your child is clear about the graph they should plot: the area values should run across the bottom, and the speed up the side. Because the time can be found for any area value (and not just for the data in the table), your child should draw a line graph rather than a bar chart.

A sudden jump

- The pictures below show a jumping toy in the 'set' position, and then in motion.

- Look at the diagrams carefully, then try to answer the questions.

1. Why does the toy stay on the ground for a short while?

_____

2. Where does the force that makes the toy jump into the air come from?

_____

3. What force will pull the toy back down to the ground?

_____

4. What would happen to the toy if the spring were made from:

much thicker metal wire? _____

much thinner metal wire? _____

5. Why is it important that the top of the toy is not too heavy?

_____

Dear Helper,

These questions will reinforce work your child has done at school looking at stored energy. They also revise key vocabulary that will be needed in future assessments. To help your child, look together at how the spring in a retractable ballpoint pen works to make the point go in and out.

Name:

Crooked pencils

You will need: a pen or pencil, a glass of water, a straight object (a ruler, a straw or another pencil), two identical coins (promise to give them back!).

● Half-fill a straight-sided glass with water. Put your straight object in the glass of water, and look at it from all the way around the glass and from above.

● In the space below, draw a picture of the object in the glass from the side and from above.

● Place the two coins on a flat surface, a few centimetres apart. Over the top of one coin, carefully place a full glass of water. Look at the two coins from above. Do they look different? Describe what you can see in the space below.

Dear Helper,

As light travels through different materials (through water as opposed to air, for example) it can give the appearance of 'bending' objects, because the light travels more slowly through some materials than others. This effect is called *refraction*. As you look at the objects in the glass of water, ask your child to describe what they can see before drawing the object. You can also observe this effect at the swimming pool – it explains why the deep end never looks as deep as the notice on the side of the pool says it is!

Who wears the glasses?

You will need to find a safe place where you can watch people walk by for this activity. Make sure you take your helper with you!

● Carry out a survey of all the people who pass you in 15 minutes. Keep a record on a tally chart of the people **with glasses** and **without glasses**.

● When you get home, draw a bar chart to show your results.

With glasses	Without glasses

Dear Helper,

We have been talking about light and lenses in class. Often the results of a survey like this are surprising: nationally, slightly more people wear glasses than do not. Ask your child to consider people who might wear contact lenses – how might they cater for these people in the survey? This activity can hopefully be used to promote the fact that people who wear glasses are not a good target for bullying – for one thing, there are more of them!

Primary colours

Primary colours are the basic colours that can be used, in different amounts, to make up all other colours. Scientists mix coloured light, but artists mix coloured paint and this is a totally different effect. Scientists and artists have different primary colours!

● Try to find out which are the **primary** colours of paint, and which are the primary colours of light.

● Once you have done this it should be a simple job to find out their **secondary** colours. Secondary colours are those made by mixing equal amounts of two primary colours. Write your findings in the space below.

An artist's primary colours are:

_____   _____   _____

An artist's secondary colours are:

_____ + _____ = _____

_____ + _____ = _____

_____ + _____ = _____

A scientist's primary colours are:

_____   _____   _____

A scientist's secondary colours are:

_____ + _____ = _____

_____ + _____ = _____

_____ + _____ = _____

Dear Helper,

We have been looking at how white light is made up by mixing different colours of light together – all the colours you see in a rainbow when combined together make white! This activity aims to show that scientists sometimes don't have the same rules that you may be familiar with in everyday life, and that can be confusing. If you go to a theatre, perhaps to see a pantomime at Christmas or to a school show, encourage your child to watch out for where the different colours of light cross or mix and make another colour.

Name:

Ding dong

● Look at this picture carefully, then answer the questions.

1. What sources of sound can you see in the picture?

_____

2. Which of the sources of the sound are the most important to hear during
a fire drill? _____

3. How does the alarm bell make a sound? _____

_____

4. How does the sound travel to your ears? _____

_____

5. Write a set of rules saying how you should behave during a fire drill.

_____

_____

Dear Helper,

This activity provides an opportunity for your child to explain where sound comes from, and how sound
travels in our environment. It also provides an opportunity to emphasise the need for sensible behaviour
during an emergency. Ask your child to describe fire drills at school – what can they hear and what must
they do?

I can't hear you!

"Don't sit so close to the TV, Sarah," said Mum. "They may be your favourite band, but you'll damage your hearing. The sound will be too loud."

Sarah wondered if her Mum was right. Were the band louder if she was nearer the TV? She doubted it! Next day at school she asked her teacher, but he said she should investigate her Mum's idea, and her prediction that Mum was wrong, to find out for herself.

● Sarah and her friend Jack did an experiment. This is the graph they drew from the results:

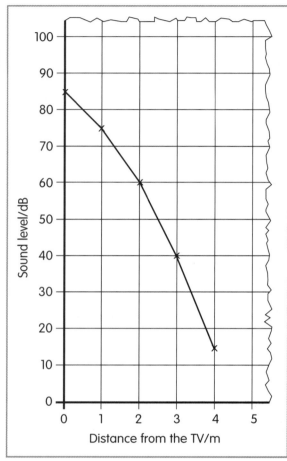

Look at Sarah's graph. Was her Mum right? _____

Describe how the sound level varies according to the distance from the TV.

_____

Describe the experiment you think Sarah and Jack did to produce this graph.

What did they need to keep the same to make sure that their test was fair?

_____

_____

Dear Helper,

This activity combines an important maths skill – interpreting data – with another important scientific skill – planning an experiment. Your child will need to think about the fairness of the test carried out. We have done similar experiments in class that your child should think back to in order to complete this activity. A fair test means changing only one thing at a time – ask your child to explain why their experiment is a fair test.

My music

- Choose your favourite piece of music. Listen to it very carefully, then try to describe it in words.

- Decide where the sounds get louder and softer, and when the sounds are at different pitches.

- What can you hear in the background, as well as the main melody? Can you guess what instruments are being played?

- What parts of them are vibrating? If there is a singer, how does their voice change?

- Some pieces of music are very complicated, so you may only be able to write about the first 30 seconds or so. Try to use some of the words in the box in some part of your description.

low	high	loud	quiet
soft	crashing	thumping	harmony

Dear Helper,

You may well be familiar with hearing the same piece of music from your child's bedroom 30 times a day, but at least this time there is a reason! This activity encourages your child to listen more carefully to the different aspects of a piece of music. They may wish to repeat this activity with different styles of music, writing a description of each and comparing the pieces. Talk about what they notice – your child does not need to write everything down. If possible, please let your child bring their music into school. Please make sure all CDs or tapes are clearly labelled.

SCHOLASTIC 87

PHOTOCOPIABLE

Name:

Is the Earth round?

It's 1492, and you are sailing across the ocean in a boat called the Santa Maria alongside your captain, Christopher Columbus. You are the ship's journalist, and are returning to the Spanish court to tell King Ferdinand and Queen Isabella about your discoveries and adventures. Your job is to write a court circular (a type of newsletter) in which you will put across your argument for the Earth being round.

● To help prepare your court circular, you will need to make a list of observations that support the idea that the Earth is round. You will also need to make a list of reasons why many people think that the Earth is flat. Remember that the year is 1492 and space travel has not happened. Use the table below to help you:

Reasons why the Earth is round	Reasons why the Earth is flat

● Study the reasons why people think that the world is flat, and try to think of an argument or observation which disproves that idea.

● Once you have done this, you can begin to write your court circular. You'll need to include details such as:

- who you are: your name, age, job
- details about the voyage: where you all thought you were going to, where you got to, the length of the journey, the conditions on board the ship
- how the journey helped you realise that the Earth is round
- observations made by others in the past that support the idea that the Earth is round
- arguments that go against the idea that the Earth is flat.

Dear Helper,

We have done a lot of work on the 'flat Earth' model, looking at evidence that supports the idea that the Earth is round. Remember that in the 15th century, people did not have the advantage of pictures from space. Talk to your child about how they know that the Earth is round. When carrying out the research, make sure your child sticks to the guidance questions.

Day and night

1. How long does it take for the Earth to spin on its axis?

_____

2. How long does it take for the Earth to orbit the Sun?

_____

3. Oliver uses a globe as a model of the Earth, and a lamp as a model of the Sun. The arrow on the picture shows the way the model of the Earth is spinning.

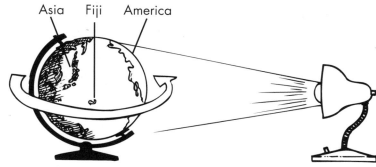

Asia   Fiji   America

When the model Earth is in the position shown in the picture about what time of day is it in:

a America? _____

b Asia? _____

c Fiji? _____

4. In the picture below, the Sun is shining early in the morning. Draw a Sun on each of the other lines to show the Sun's height at the times given.

5. Why does the Sun appear to move across the sky?_____

_____

_____

Dear Helper,

Not only does the Earth rotate around the Sun (which causes the seasons), but the Earth spins on its own axis as well, completing one revolution every 24 hours (which is what causes day and night). We have been looking at what causes day and night in class. Help your child answer question 4 by encouraging them to look for the position of the Sun at various times during the day.

⚠ **DO NOT let them look directly at the Sun!**

Daylight and darkness

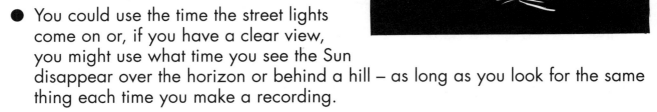

● Twice a week for the next month, use this sheet to record the time the Sun sets.

● To make your recordings reliable you will have to decide how you know when the Sun has set.

● You could use the time the street lights come on or, if you have a clear view, you might use what time you see the Sun disappear over the horizon or behind a hill – as long as you look for the same thing each time you make a recording.

● Record your observations in the table below, on days that are not dark, cloudy, misty or foggy.

Day	Date	Sunset time

These results need to be brought back to school on: _____

Dear Helper,

This is an extension of work your child has done in class on the causes of day and night. To reinforce this work, ask your child why it is daylight in Australia when it is dark in the UK. These observations need to be made twice a week over the course of a month – reminders may be needed!

Name:

Making a sundial

You will need: A stick or broom about 1 metre in length, 12 stones, 12 small sticks or pencils, string, sticky labels (or paper and sticky tape), scissors, a pen, paper, a sunny garden or yard.

What to do:

1. In an open area that won't be affected by other shadows, push the big stick into the ground or prop it up between two bricks.

2. Tie a piece of string around the bottom of the stick and line it up with the shadow cast by the stick. Cut the string where the shadow ends, so the string is the same length as the shadow.

3. Tie the end of the string to a small stick or stone and use this to mark the point where the shadow ends. Label this string with the time it represents.

If you can't find any string just place a stone at the end of the shadow and label the stone with the time at which that shadow was formed.

4. Repeat this every hour, marking the shadow and labelling the time that it represents. Try making recordings for at least 8 hours (starting at, say, 9am and continuing until 5pm).

5. Once you have made all the recordings, draw the pattern made by the shadows on the back of this sheet. Write the time on each shadow line.

6. Go out tomorrow and look at your sundial again. Can you tell the time with it?

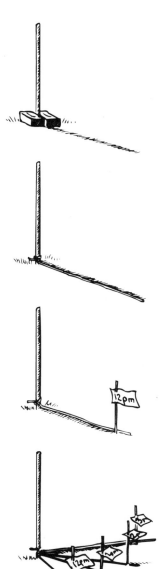

Dear Helper,

Help your child make a sundial by following the instructions above. You will need to find an open space in which to create your sundial so it won't be disturbed or affected by other shadows – parks are not a good place! Once you've completed the readings, encourage your child to tell the time with the sundial.

⚠ **Make sure that there are no young children or animals nearby who might hurt themselves on the sticks or trip up on the string. If this is the case, use the stones instead, but make sure young children don't put these into their mouth! Don't let your child out to do this activity on their own – supervision is needed at all times.**

PHOTOCOPIABLE

Name:

Day-lengths

● This table shows each of the planets in our Solar System and the time it takes for each planet to spin once on its axis (measured in Earth time).

Planet	Time to spin on axis/Earth hours	Time taken to spin on axis/Earth days
Mercury	1400	
Venus	5800	
Earth	24	
Mars	25	
Jupiter	10	less than 1
Saturn	10	less than 1
Uranus	17	less than 1
Neptune	16	less than 1
Pluto	150	

1. How many hours does it take the Earth to spin once on its axis?

_____

2. If one Earth day is 24 hours, work out how many Earth days it takes the other planets in the table to spin once on their axis. Write your answers in the table. Some have been done for you.

3. Draw a bar chart to show how long it takes Earth, Mars, Jupiter, Saturn, Uranus and Neptune to spin once on their axis. Use the data from the table.

Dear Helper,

This activity revises your child's maths skills, which they will have practised in class. Remind your child that when drawing a graph they must use a well-sharpened pencil and take care with regard to accuracy and neatness. It is important to label the axes and give the graph a title.

PHOTOCOPIABLE

The four seasons

1. Look at these four pictures:

Write the name of the season shown under each picture.

2. Why do we have different seasons? _____

_____

3. On this diagram below, the Earth is shown in each of the four seasons. If you were at the point marked **X**, write whether it would be spring, summer, autumn or winter next to each of the positions.

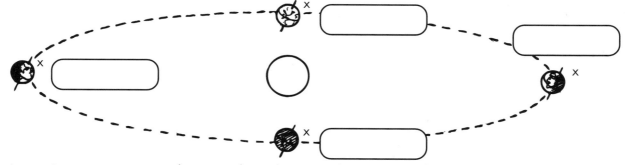

4. Imagine you are at the North Pole. At what time of year is it:

a daylight for 24 hours? _____

b darkness for 24 hours? _____

Dear Helper,

The seasons are caused by the movement of the Earth around the Sun. At some points the UK is tilted towards the Sun (this is summer), and at others we are tilted away (winter). Challenge your child to name some countries in the northern hemisphere and some in the southern hemisphere, and ask them to think about what season it will be in Australia when it is winter in the UK.

The Earth's 23½° tilt

1. Use the following words to complete the passage.

tilted	**away**	**lower**	**Earth**
towards	**fewer**	**higher**	**longer**

As the _____ moves around the Sun, it is always _____

the same way. In summer, the UK is tilted _____ the Sun. The Sun

appears to be _____ in the sky, and daylight lasts

_____. In winter the UK is tilted _____ from

the Sun. The Sun is _____ in the sky, and there are

_____ hours of daylight.

2. Using your scientific knowledge and your imagination, describe how our lives would be different if the Earth's axis was not tilted at all.

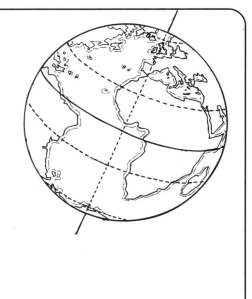

Dear Helper,

To help with question 2, you could perhaps use a globe or football as a model. Ask your child to show you how the Earth is tilted as it orbits the Sun, and how this affects us in the UK. They can then un-tilt the Earth to try to work out what changes would occur if the Earth's axis were not tilted.

Family fitness plan

Most of us spend far too much time in front of the TV or sitting at a computer, and being inactive for long periods can have serious side-effects. Doctors recommend that we do at least 20 minutes exercise each day. When you are young this is not so difficult, but for older people in your family finding time to exercise is not always so easy.

- Talk to your family and together come up with some activities that you can all take part in. Keeping fit should be fun – playing football in the park, taking the dog on an extra long walk or going swimming as a family all bring enormous health benefits.

- Make time to keep fit! Agree a time in the day when the TV is switched off for an hour, and use this time to get active!

- The chart below is to help you plan some activities.

Family member	Exercise they do now	New activity	When?	How frequently?

Dear Helper,

Heart disease is one of the most common killers of adults in Britain today. A major cause of heart disease is an inactive lifestyle. Many adults develop the first signs of heart disease in early life, but being physically active can help to reduce the chances of suffering. Encouraging your child to maintain an active lifestyle could help safeguard them from an unnecessary and crippling disease. Share ideas to help keep activity fun!

NEW BEGINNINGS (OURSELVES) UNIT 1

Lifestyle muddle

● Look at the list below. Try to match the activities to the sentences that describe their effects.

Eating too much sugar and fat	Are prescribed by a doctor, and help to fight illness.
Taking exercise	Inhaling their vapours can cause brain and liver damage.
Fibre	Can cause heart and respiratory diseases, including cancer.
Recreational drugs	UV light causes skin cancer. Can be prevented by covering up or wearing sunscreen.
Sunbathing	Can cause hearing loss.
Cigarettes	These are not prescribed by a doctor. Their unknown contents may cause serious damage, even death.
Loud music	Strengthens the heart and other muscles. Improves the circulation and general fitness.
Medicinal drugs	Found in vegetables, fruit and cereals. Helps to keep our digestive system healthy.
Solvents	Can lead to obesity (being seriously overweight) and put stress on the heart. It can lead to other medical complications.

Dear Helper,

We have been discussing health and health risks at school. Talk with your child about some of these activities. Ask them to tell you the benefits or risks of each activity, and ask them to think about their own lifestyle. Perhaps as a family you could think of targets to keep yourselves healthy, and have fun keeping fit!

Staying alive!

Serious diseases are still common in many parts of the world. In poorer countries many children die before they reach the age of six.

● This table shows what happened to 500 children in three different countries.

Cause of death	Uganda	Bolivia	Sweden
Malaria	128	25 ☐ %	0
Tuberculosis (called 'TB')	89	50 ☐ %	0
Measles	67	75 ☐ %	0
Other causes	179	125 ☐ %	27
TOTAL	500	500	500
Total died (A)			
Total survived (B)			

1. a Work out how many of the children died in each country. Fill in the answers in row (A) of the table.

b Use your answers to question 1a to find out how many children in each country survived. Write these answers in row (B) of the table.

2. Work out the percentage of children in Bolivia that died of each disease. Write your answers in the boxes in the table.

3. These pie charts show how many children died in Uganda and Sweden.

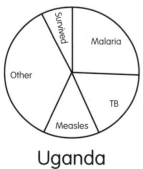

Uganda

Bolivia

Sweden

● Use the figures in the table to fill in the pie chart for Bolivia.

Dear Helper,

We have been thinking about how people are born, grow and die – the life cycle. This activity will help your child recap their data-handling skills, using addition and subtraction, and will assess their ability to work out percentages to compare data from three different countries. Look at the data; point out how few children die young in a rich country such as Sweden compared to the other countries – why do they think this is? Challenge your child to find Uganda and Bolivia on a map.

PHOTOCOPIABLE

Name:

Piggy-back tadpoles

● Read through this text and then answer the questions below.

The strawberry dart frog lives in the rainforests of South America. The female lays her eggs in pools of water that collect in the big green leaves of plants growing on the forest floor, and the eggs are fertilised by the males.

Once the fertilised eggs have developed into tadpoles, the mother frog encourages one to wriggle on to her back, which is sticky. She gives the tadpole a piggy-back ride and climbs up into the branches of the trees. Here she seeks out a pool of rainwater in the leaves of a bromeliad (a special plant that grows on the branches of rainforest trees). The mother frog releases the tadpole into the water and lays another, edible, unfertilised egg in the water.

She then climbs down to pick up her next tadpole, and sets off with it to find another leaf pond. In this clever way, the strawberry dart frog spreads her young through the forest and increases their chance of surviving to become adults.

The tadpoles grow into frogs with vivid, strawberry-red skin. Their skin is so toxic that just by brushing the tips of their blow-darts against the frog, people living in the forest can produce deadly poisonous weapons.

1. Why do you think the female frog leaves an unfertilised egg with the tadpole? _____

2. Why do you think the tadpoles have more chance of surviving if they are spread out? _____

3. The female British common frog lays hundreds of eggs. Only a few of the tadpoles survive to become adult frogs. Why do you think this is?

_____

4. Red is a poor colour for camouflage. Why do you think the strawberry dart frog is such a bright colour? _____

Dear Helper,

Your child has been looking at the way human parents provide care for their young. This homework links to work on classifying vertebrates and to an appreciation of the fact that living things have a range of strategies for caring for their young. Help your child read through the text if necessary, and ask them to look up any words they don't understand. You might like to find South America on a world map together too.

◣ SCHOLASTIC

100 SCIENCE HOMEWORK ACTIVITIES ● YEAR 6

Reproduction crossword

Clues across

2 The _____ sac contains fluid to protect the baby.

5 A fertilized egg.

8 Another word for the mother's womb.

9 Part that helps a sperm cell to swim.

11 Attaches the umbilical cord to the mother.

13 Eggs are released from here.

14 Male sex organs that produce sperm.

15 Carries nutrients and oxygen to the baby.

Clues down

1 Name for a baby in the womb in its first few weeks.

3 Tube that carries eggs from the ovary to the uterus.

4 Process when an egg and sperm cell join.

6 Cycle lasting 28 days in the female.

7 When a mother is expecting a baby, she is said to be this.

10 Male sex organ.

12 Name for the developing baby in the womb after its first few weeks.

Dear Helper,

The answers to these questions revise the work your child has been covering in class about pregnancy and birth. It may be useful to have a reference book or dictionary handy as you complete this crossword. Like all crosswords, some of these answers are easier than others! If you wish to find out more about the topics covered by your child, or our sex education policy, contact your child's teacher for details.

Answers: Across – 2 amniotic; 5 zygote; 8 uterus; 9 tail; 11 placenta; 13 ovary; 14 testes; 15 umbilical cord. Down – 1 embryo; 3 oviduct; 4 fertilisation; 6 menstrual; 7 pregnant; 10 penis; 12 foetus.

Animal parents

You will need: scissors and sticky tape or glue.

Vertebrates (animals with backbones) can be divided into five groups: mammals, reptiles, amphibians, fish and birds.

● Which group do humans belong to? Write **humans** in the correct space in the table below.

● Think about the special features this group has. Try to name some other examples of animals that belong to this group. Try to name two examples of animals in each of the other groups.

● Look at the statements at the bottom of the page. They describe the habits of parents in each group of animals. Work out which habits you think fit the parents in each group and write the letter in the last column.

Group	Animal examples	Parents' habits
Fish		
Reptiles		
Mammals		
Birds		
Amphibians		

● Choose from these:

A Parents have dry, scaly skin, and lay hard-shelled eggs on land.

B Parents have feathers and lay eggs on land.

C Parents have gills and scales, and lay eggs in water.

D Parents have hair or fur and provide milk for their young.

E Parents live in and out of the water, have soft skin, and lay eggs in water.

Dear Helper,

If possible, have reference books handy for this activity – or visit your local library together – so your child can find out a little more about each of the vertebrate groups. Test your child once they have finished their work to see how much they can remember about each of the five groups. Challenge them to find out which group the duck-billed platypus belongs to. What is unusual about this strange creature as a parent in that group? (The duck-billed platypus is a mammal that lays eggs, but provides milk for its young.)

Coral reefs

Coral reefs are rich and colourful sea (marine) habitats. They provide food, shelter and a breeding ground for a huge variety of life.

_____

_____

_____

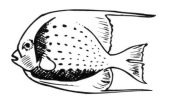

_____

_____

_____

_____

_____

_____

● The organisms (living things) shown above are just a few you might find living near a reef. Some of them are plants, some are invertebrates (they don't have a backbone) and some are vertebrates (animals with a backbone).

● Underneath each organism, write the name of the group you think it belongs to.

1. Which coral reef is found off the north-east coast of Australia?

_____

2. What threats are there to coral reefs? _____

_____

Dear Helper,

Being able to classify living things according to their characteristics is a useful skill, and one which may be required in the national tests. We have been looking at how scientists classify creatures in class. We have talked mostly about local habitats, but this homework challenges your child to apply their knowledge to a less familiar setting. If you have access to the Internet, try looking at www.bbc.co.uk/nature, which provides a fun and fascinating introduction to the environment. Ask your child to tell you how they are deciding which creatures belong to which groups.

VARIATION (ANIMALS & PLANTS) UNIT 2

Ancient animals

● Imagine you have travelled back in time. Around you is a world full of prehistoric vertebrate animals.

● Read the description of each animal below, and try to identify its correct group. Choose from:

reptile	amphibian	mammal	fish	bird

Name	Description	Group
Stegosaurus	Stegosaurus is a heavy plant-eating animal with dry scaly skin. Bony plates stick out from its neck and run down its back to the large spikes on its tail. It lays hard-shelled eggs on land.	
Machairodus	This animal looks like a big cat with claws and huge sabre teeth, ideal for ripping the hide of a mammoth. It has fur and provides milk for its cubs.	
Archaeopteryx	This prehistoric animal glides between the ancient trees. Unlike other gliders, it has wings covered in feathers. It lays eggs with hard shells.	
Dinichthys	This 12-metre long sea creature has an armoured head, fins and gills. It hunts other sea animals. It lays eggs in the water.	
Eryops	Eryops lays its eggs in water. It looks like a crocodile but has smooth damp skin. It breathes air, and can live in or out of the water.	

What is a vertebrate? _____

_____

Dear Helper,

We have been looking at what features of animals make each of the five vertebrate groups (reptile, amphibian, mammal, fish and bird) distinct. This exercise will help to see if your child understands the main differences. To help them complete this activity, encourage your child to think of modern examples of the animals listed above, and to think about what these creatures have in common with the ancient animals? To start you off, a frog is an amphibian.

A guide to my family and friends 1

- You've seen how to make branching keys in school. Now it's your turn to make a guide to your family and friends.

- Choose eight people (four male and four female), and think about how they could be divided up, other than by their sex.

- Remember to use **physical** features that other people could recognise by using your key.

- Try not to insult your auntie by describing her as very old and wrinkly with a big nose.

- **TIP:** Look at eye and hair colour. Do they wear glasses or have brown hair? Your key might look like this:

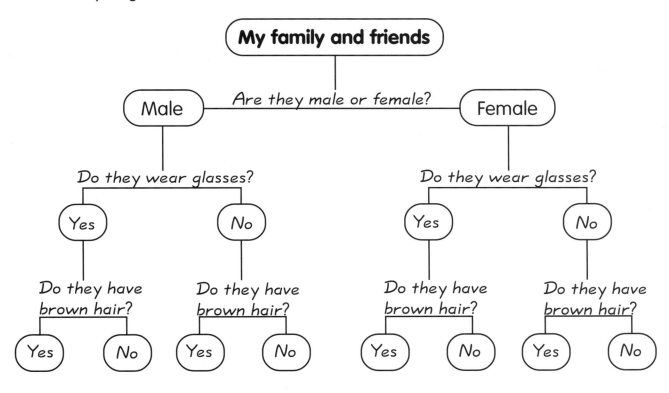

- Once you've completed your key, draw a picture (or use a photo – but remember to ask permission!) of each of the people in your key.

- Bring them in to school, along with your finished key. Give each picture a letter and see your friends can match the correct name to each face using your key.

Dear Helper,

Knowing how to use a key is a useful skill for your child to develop, and is an aspect of science work that may come up in your child's national tests. Help your child to organise their thoughts when dividing your friends and family into groups – each question must have only two possible answers, and whatever is written on one of the branches labelled A must also be written across the other labelled A. This goes for the other branches too. So, if one branch C is labelled *has a beard*, all the other Cs must be the same.

A guide to my family and friends 2

VARIATION **ANIMALS & PLANTS** **UNIT 2**

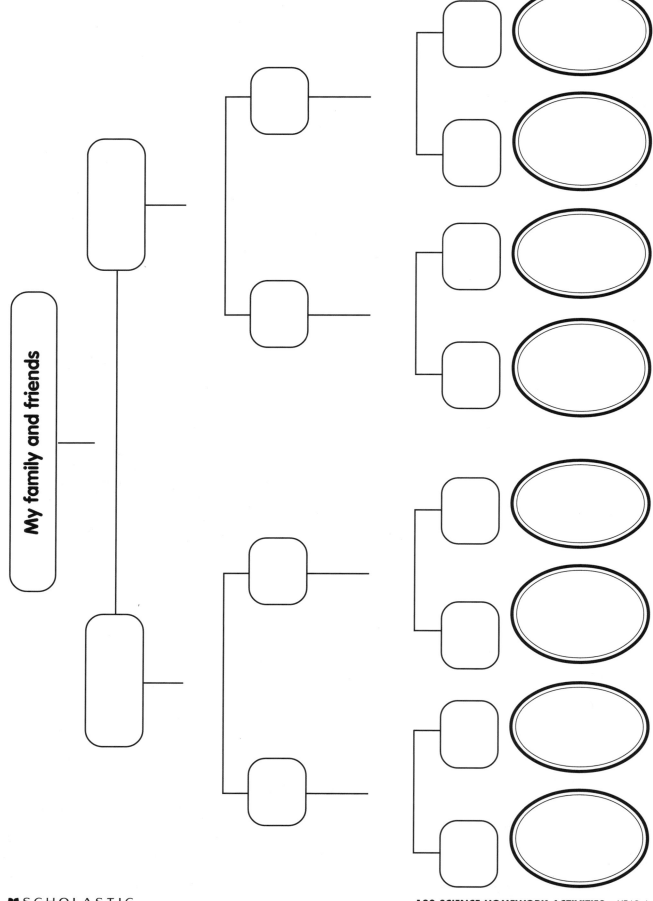

My family and friends

Variations

● Read these extracts, then answer the questions below.

The peppered moth comes in two forms, one with lightly speckled wings and one with darker wings. They rest on tree trunks during the day, hiding from predators such as birds. During the Industrial Revolution many trees became covered in dark soot from factories, and the darker form of the peppered moth became more common.

1. What does **variation** mean? _____

2. What was the main type of variation in the peppered moth population?

_____

3. Why do you think more dark peppered moths survived on sooty trees?

_____

4. Why was it good for the peppered moths to have variations in colour?

_____

A group of killer whales is called a pod. Whales, like peppered moths, have variations in their appearance. Look closely at this pod of killer whales, then try to answer the questions.

5. Name a discontinuous variation that could be used to group the whales.

_____

6. Name a variation which is gradual (a continuous variation).

_____

Dear Helper,

Ask your child to explain *variation* to you. While we are all the same in that we are humans, we all have many differences too – height, blood group, gender and so on. To help your child understand variation, next time you are out, look for different breeds of the same type of animal (for example, dogs). Talk about how they are the same and how they vary.

Plants in space

● Read the passage below and then answer the questions.

Take a deep breath. Our bodies need Factor X from the air in our lungs. Without it we would not survive. Billions of years ago the Earth's atmosphere had very little Factor X in it. When the first plants floating in the oceans began photosynthesising, they took in the gas carbon dioxide and began to produce Factor X. Thanks to plants, the Earth's atmosphere is now 21% Factor X and only 0.03% carbon dioxide.

This gave scientists the idea of using plants to provide the Factor X that astronauts need on long space flights. In an experiment, a volunteer was sealed in a large airtight chamber with thousands of barley seedlings. The seedlings were grown under bright lamps and the roots trailed in water that had essential nutrients added to it.

The experiment was successful. Fans circulated the air in the chamber and while the volunteer exercised, worked or slept the plants took in carbon dioxide and produced the Factor X they needed.

1. Factor X is a gas with a proper science name. Which gas is it?

_____

2. Which word describes the special way plants make food in their leaves?

_____

3. Name four things plants need to grow well. (Clue: the lamps provided two of them!) _____

4. Apart from producing Factor X, why else might plants be useful in space?

_____

5. What disadvantages could there be if you relied completely on plants for Factor X in a spaceship? _____

Dear Helper,

Wherever green plants grow, they have similar needs in order to remain healthy, and carry out similar life processes. Ask your child about the experiments they have done in class about growing healthy plants. Challenge your child to think about how farmers try to grow the best possible crops – what do they provide?

Name:

Growing tomatoes

● Zaidi has been helping his mum grow tomatoes in bags of soil in their garden. They put three tomato plants into each of four bags. As an experiment, Zaidi added different amounts of fertiliser (called **nitrate**) to three of the bags.

● With his mum, he weighed the tomatoes produced by each plant. His results are in the table below.

	Plant 1	Plant 2	Plant 3	Average
Bag A	0.5kg	1.0kg	1.5kg	_____ kg
Bag B	2.0kg	3.0 kg	1.0kg	_____ kg
Bag C	2.5kg	5.0kg	1.5kg	_____ kg
Bag D	0.3kg	1.0 kg	0.2kg	_____ kg

Bag A contained no added nitrate

Bag B had 10g of nitrate added

Bag C had 20g of nitrate added

Bag D had 30g of nitrate added

1. Work out the average weight of tomatoes made by the plants in each bag.

2. Draw a graph of Zaidi's results on these axes:

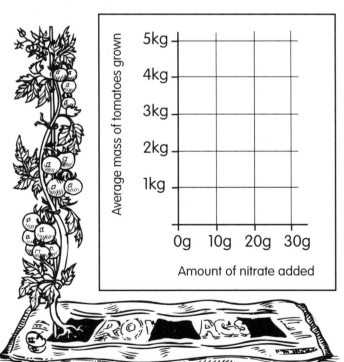

Average mass of tomatoes grown

5kg
4kg
3kg
2kg
1kg

0g   10g   20g   30g

Amount of nitrate added

3. Using the information above, how much nitrate fertiliser would you choose to add to a bag of soil like the ones Zaidi and his mum used? Give a reason for your answer.

4. Why do you think it was useful to have a bag with no fertiliser added to it?

5. If a farmer wanted to work out from Zaidi's experiment how much fertiliser to add to a field, what other information would it be useful for him (or her) to know about the experiment?

Dear Helper,

Help your child read through these questions, and help them with their calculations if necessary – without using a calculator! Encourage your child to write down their working out as they do this. To find the average weight of the three tomatoes in each bag, add the results for each bag together and divide by 3.

PHOTOCOPIABLE

Name:

Garden food webs

Mary's dad was pleased when he spotted a hedgehog in the garden one night. "Just in time!" he said, explaining, "they will save our lettuces from the slugs."

Once, Mary saw a fox crossing the lane when she was on her way to school. She was amazed when her mum told her that foxes sometimes killed hedgehogs – she had always thought that the hedgehog's spines would protect it from its enemies.

● Mary tried to make a food chain for her garden.

1. Help her to fill in the blanks.

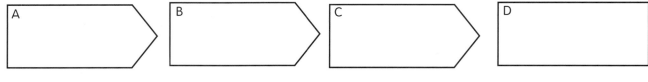

2. Which of these is a producer? _____

3. How does the producer get the energy it needs? _____

_____

4. Which of these are predators? _____

5. What do you think will happen to the number of foxes if it is a bad year for growing lettuces? _____

6. What do the arrows in a food chain show? _____

_____

Dear Helper,

A food chain shows feeding relationships – what eats what. They always start with a producer (a plant), and there may be several consumers that create a food chain. This activity will help your child gain confidence in organising living things into food chains. Read through this sheet together. If they struggle with the questions ask them to think about the work they have done on food chains in school. Ask your child: *What do all plants need in order to live?* Remind them that producers (in this case, lettuce) are always the first step in a food chain.

Adaptations for survival

Barn owls are hunters of the night. Even if you have never seen one, it's possible you have heard its eerie screech. Barn owls often nest in old buildings and barns. They have wide downy wings, which give them a silent flight. They spot prey using their big eyes, which are very sensitive even in dim light. They have powerful legs and sharp talons for seizing mice and voles, and a hooked beak helps them to tear at the flesh of animals, although as adults they often swallow their prey whole.

● Using this information, give a reason for each of the owl's adaptations.

The Venus fly-trap grows in marshy soils (bogs) that contain little nitrate. Nitrate is an important ingredient in fertilisers, and without it plants have difficulty making proteins. To gain extra nutrients, the Venus fly-trap uses its leaves as clever traps. The hinged, open leaves are covered in hairs that are sensitive to touch. When a small bug crosses the leaf it springs a trap. The overlapping spines on the edge of the leaf trap the victim and the leaf produces chemicals that digest the body, providing the plant with all the protein it needs. Other leaves on the plant make sugars, and the roots anchor the plant to the soil, helping it to take up water.

● Explain how a Venus fly-trap is adapted to life in a marshy soil.

Dear Helper,

We have been looking at how living things adapt to where they live (their habitat). Help your child to read through these two descriptions and pick out the special features which help each survive in its habitat.

Name:

Lichens and air pollution

Lichens (usually pronounced 'lie-kens') grow on rocks, walls and trees. There are thousands of different types of lichen. They are made of an algae and fungi living together – the algae makes the food and the fungi protects it and stops it drying out. Some lichens grow like flat grey-green circles and can be seen on old gravestones and walls. Some grow feathery fronds.

Lichens have no roots, and so take in water and gases through their surface. Because of this they are very sensitive to pollution.

A line transect was carried out either side of a motorway. Every 500m a quadrat was placed against a nearby stone wall, and the number of lichens found in the quadrat was recorded. Three types of lichen were found – A, B and C. The results are shown below.

Lichens	1500m	1000m	500m	0m		0m	500m	1000m	1500m
A	5	4	3	2	MOTORWAY	1	2	4	4
B	6	4	2	0		0	0	5	5
C	5	2	1	0		0	0	0	1

Usual wind direction: south-west

1. There is air pollution is this area. What do you think causes it?

_____

2. Which type of lichen – A, B or C – seems least sensitive to pollution? Why do you think this? _____

3. Which lichen would be a useful indicator of clean air? _____

4. Why do you think there are fewer lichens on the eastern side of the motorway? _____

Dear Helper,

Your child will have already used a metal frame (called a quadrat) to carry out a study of the plants in the school grounds. Plants are often used by scientists to measure air quality because of their sensitivity to pollution. This exercise will help your child understand and interpret results presented in tables.

■SCHOLASTIC **100 SCIENCE HOMEWORK ACTIVITIES** • YEAR 6

The soil where I live

You will need: a sample of soil (no more than enough to half-fill a margarine tub). Make sure you ask your helper if you can dig up soil and bring it into the house! If it is difficult for you to get hold of soil from home, perhaps you could take some home in a tub from school.

> ⚠ **You will need to handle the soil, so it's important to check first that you have no open cuts on your fingers (cover them with a plaster). Check the soil is free of any sharp objects.**

● To prevent making a mess, clear space on a table and cover it in old newspaper before you start.

● What type of soil have you got?

● Moisten the soil with a little water (but not so much that it turns slushy). Now try following these steps.

  • Take a heaped handful of soil and remove any stones you can see.
  • Use your hands to squeeze the soil into a ball shape.
  • Now roll the soil into a sausage shape.
  • See if you can bend the ends of the sausage together to make a big ring.

● Look at the chart below and see what kind of soil you have:

Sandy soil	Loam (a mix)	Clay soil
Begins to crumble as you roll it into a sausage. Difficult to mould.	Rolls into a sausage but begins to crack when bent into a ring shape. Moulds fairly well.	Rolls easily into a sausage and into a ring. Moulds and holds its shape easily.

1. Which type of soil is yours most like? _____

2. Sandy soils drain very quickly and clay soils can hold puddles for a long time. Why do you think most gardeners prefer a loam soil? _____
_____

3. How are worms specially adapted for life in the soil? Why are they good for the soil? Try to find out! _____
_____

Dear Helper,

This activity will help your child (any maybe yourself!) discover more about the world around them, and encourages independent research. If you have trouble obtaining a reasonable sample of soil, your child may be able to get some from school. If possible, visit the local library with your child, or look on a CD-ROM or use the Internet, to find out more about soil. (Sorry about the mess.)

Name:

Preventing decay

George Mallory and Andrew Irvine are famous for their attempt to climb Mount Everest in 1924. Neither survived – both died on the mountain not far from the summit. George Mallory's body was discovered 75 years later, still preserved, lying among rocks on the icy slopes.

Sabre-toothed tigers have been extinct for many thousands of years, but some bodies have been pulled out of ancient tar pits, their flesh and bones preserved by chemicals.

The Ancient Egyptians mummified bodies by removing the moist organs, packing the flesh with dry herbs and spices and wrapping the bodies in dry bandages. This technique preserved the flesh for thousands of years.

Food sealed in unbroken tins will last for many years, but once the seal is broken the food will quickly decay.

Decay is caused by micro-organisms, which include bacteria and fungi (moulds). By breaking down materials they help to release nutrients that can then be used again by other organisms.

1. Why do you think Mallory's body had not decayed? _____

_____

2. Why do you think the bodies of sabre-toothed tigers have not decomposed in tar pits? _____

3. What was important in preparing the body of a mummy so it would not decompose? _____

4. Why do you think it is important that cans for storing food are airtight?

_____

5. Put a circle around the conditions that you think would help bacteria and fungi cause something to decay quickly: **moist   dry   cold   warm**

without air   with air   with toxic chemicals   without toxic chemicals

Dear Helper,

This exercise will give your child practice of comprehension-type questions. It should also help your child revise what they have learned about microbes and decay at school, and maybe encourage them to find out more! If necessary, read through the extracts with your child, and talk about why they think decay has been prevented in each case.

Useful microbes

● Read the sentences, and choose the letter that shows the use made of the microbe. Write this letter in the matching box.

Microbes

1. Fungi and bacteria are rich in protein. ☐

2. Yeast produces carbon dioxide gas. ☐

3. Mould grows on curdled milk adding flavour as it matures. ☐

4. Yeast turns sugar into alcohol in the fermenting process. ☐

5. Fungi and bacteria break down waste materials. ☐

6. Bacteria cause milk to curdle. This can then be flavoured. ☐

Uses

A Rotting down compost and sewage treatment

B Helping bread to rise in baking

C Making wine and beer

D Making meat substitute

E Producing yoghurt

F Making cheese

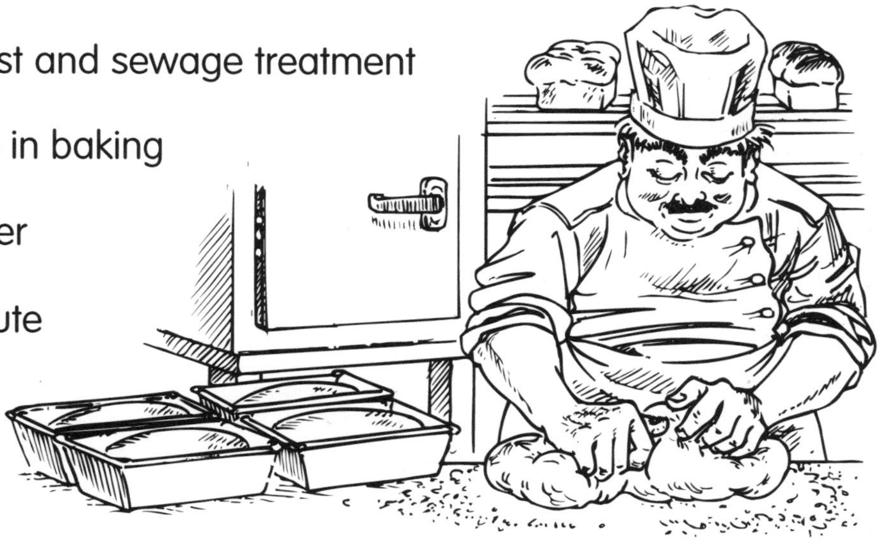

● Think about the number of ways you have used microbes this week and write them down on the back of this sheet.

Dear Helper,

At school, we have been looking at the importance of microbes (in particular, bacteria and fungi) in our everyday lives. Encourage your child not to rush this exercise, but to explain why they have made each choice. If you have reference books or access to the Internet, encourage them to find out more ways in which microbes are useful!

Soluble or insoluble?

You will need: teaspoons, cups, warm water, a small bowl, and about a teaspoonful of: instant coffee*, flour*, sugar*, bicarbonate of soda, icing sugar, custard powder, gravy granules, and 'hundreds and thousands'. Remember to ask permission before taking anything from food cupboards.

● Try this experiment. Test as many of the foods listed above as possible. Don't worry if you don't have some of the things; if you can only do the ones marked with a *, that's OK.

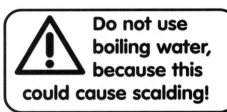

Do not use boiling water, because this could cause scalding!

1. Pour a cupful of warm water into a small bowl.

2. Take a level teaspoon of flour, add it to the tepid water and stir.

3. Does the flour dissolve in the water? Fill in the chart below.

4. Wash and dry the equipment you have just used.

5. Repeat the activity, using the same equipment, for all the other foods. Remember to wash up each time!

Try to use the same amount of each ingredient and water each time. Why should you do this? _____

Ingredient	Dissolves in water	Does not dissolve in water

1. Which of the ingredients you tested did not dissolve in water?

2. How could you separate one of the insoluble ingredients from water? You may wish to use diagrams to help you describe it.

Dear Helper,

We have been mixing and dissolving lots of *materials* at school. Ask your child to explain what a material is, then help them to try out this experiment at home using small amounts of some common foodstuffs. (Don't worry, you shouldn't need to buy anything special!) If you don't have all the suggested items, feel free to try some other ingredients. As an extra challenge, find out what happens to olive oil or vegetable oil in water.

Ice and water

● Georgina pours a glass of water and puts some ice in to keep it cool. After a while, droplets appear on the outside of the glass. She leaves the drink on the table for an hour.

1. What is the name of the process that causes the droplets to form on the outside of the glass?

2. Where does the liquid that makes these droplets come from?

3. What will happen to the ice in the glass?

● At school, Tom puts some ice in a beaker of water. He measures the temperature of the water with a thermometer every 20 minutes, and plots the results as a graph.

4. **a** What is the temperature of the water at the start of the experiment?

   b Why is the water at this temperature ?

5. What is the temperature of the water after 20 minutes?

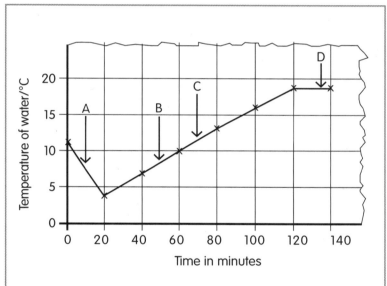

6. **a** After 20 minutes is the temperature of the water lower, higher, or the same as it was at the start?

   b Why do you think this?

7. At which point (**a**, **b**, **c** or **d**) will there be ice in the beaker?

8. What is the temperature of the air in the room Tom is working in?

Dear Helper,

We have been looking at how solids change into liquids and vice versa. Ask your child to explain what happens to an ice lolly when it is left in a bowl in the kitchen, and ask them what this process is called. Challenge them to explain what would happen if you put the liquid back into the freezer and what this process is called.

REVERSIBLE AND NON-REVERSIBLE CHANGES (MATERIALS (UNIT 4

Name:

The dissolving sugar experiment 1

● Isabella and Robert carried out an experiment.

● They had four bowls of water. Each was a different temperature:

4°C   20°C   30°C   40°C

● They added some sugar to each bowl, and stirred it with a spoon, and timed how long it took for the sugar to dissolve.

● These are their results:

Temperature of water/°C	Time taken for sugar to dissolve/s
4	135
20	77
30	55
40	23

1. How long did it take the sugar to dissolve in water that was at 4°C?

2. How long did it take the sugar to dissolve in water that was at 40°C?

3. If Isabella and Robert had used water at 50°C, what do you think they would have found about how fast the sugar dissolved?

4. Why do you think their teacher did not let them use water with a temperature of 50°C or above?

5. How might Isabella and Robert have made their experiment fair?

Dear Helper,

We have been carrying out some experiments to see how the temperature of water affects the speed at which sugar and salt dissolve. Before we tried this, your child had to make a prediction about what they thought would happen – for example *The hotter the water, the quicker the sugar will dissolve*. They also had to consider how they would make the test fair, such as using the same amount of water and sugar for each bowl, thus changing only the temperature of the water.

Ask your child about the experiment they did in science: *How did you make the experiment fair? What did you find out from your results? Were your results what you expected?*

PHOTOCOPIABLE

SCHOLASTIC   **100 SCIENCE HOMEWORK ACTIVITIES** ● YEAR 6

The dissolving sugar experiment 2

● These are Isabella and Robert's results for their experiment:

Temperature of water/°C	Time taken for sugar to dissolve/s
4	135
20	77
30	55
40	23

● Using the axes below, draw a line graph of Isabella and Robert's results.

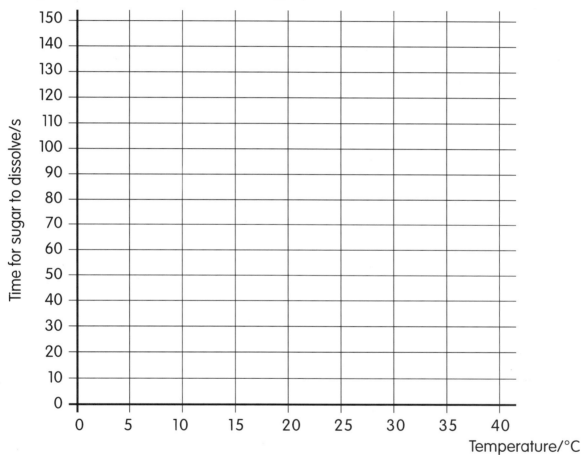

● Use the graph to answer the following questions:

1. How long would it take the sugar to dissolve in water at 25°C?

2. If it took 100 seconds for the sugar to dissolve in the water, what is the temperature of the water?

Dear Helper,

This activity follows on from the last homework your child did on dissolving sugar in water at different temperatures. This activity focuses on using and interpreting the experiment's results. It would help if you could ask your child questions about the experiment; about what they did and what the results showed them. When drawing the graph, remind them to use a well-sharpened pencil, and to pay attention to neatness, clarity and accuracy.

Name:

Mixing materials

● Nicolas and Alex were mixing materials. They mixed different materials in four clear plastic bags, and tied the top of each bag. These are their observations:

Bag	Mixture	Observations	
A	Blue copper sulphate crystals and water	Water turns blue and you cannot see the copper sulphate crystals after a while.	
B	Vegetable oil and water	The oil floats on top of the water.	
C	Bicarbonate of soda and vinegar	Lots of fizzing. The mixture looks frothy, and the bag puffs up.	

1. Write down the names of the three liquids that the children used.

_____

2. Look at the results. In one bag, dissolving was the only change. Which bag was this? _____

3. The mixture in bag C fizzed and the bag puffed up. Why did bag C puff up?

_____

4. Two of the mixtures can be separated to get the starting materials back again. One of the bags contains a mixture that can't be separated back into the starting materials. Which bag contains the substance that cannot be separated back into the original starting materials?

Dear Helper,

Your child has been looking at mixtures and how to separate them. When some materials are mixed together changes occur that cannot be reversed and something new is formed – this new substance cannot be separated back into its original materials. Ask your child what they noticed in class when they added Andrew's Liver Salts to water and what this showed. Do they think that new substances were formed, and could these new substances be turned back into Andrew's Liver Salts and water again? (Carbon dioxide was formed, and in this case the materials cannot be separated.)

Burning

- Ciara holds different materials in tongs over the flame of a candle.

- She writes down what happens to each material. Look at her notebook below.

Material	Observation
Chocolate	Melts and bubbles, then smokes and turns black.
Wax	Drips, smokes and flames. Nothing is left on the tongs.
Brick	Black coating. No other change.
Cotton	Flames and grey ash left.
Steel paper clip	Black coating. No other change.
Paper	Flames, and a thin black material left.
Wooden stick	_____

1. Complete the notebook to show what will happen to the wooden stick in the flame.

2. Which two materials melted and then burned?

3. Before burning, Ciara weighed the wooden stick. It had a mass of 2g. After burning the wooden stick, Ciara found the mass of what was left. What would she have noticed about the mass of what was left after burning?

4. Does burning cause an irreversible or reversible change?

Dear Helper,

Burning is an *irreversible change* – once something is burned, you cannot get the original material back. Ask your child what was left after burning a wooden splint – could they put these leftovers back together to make another wooden splint?

⚠ **When carrying out this homework, do not let your child try and carry out the activities that the imaginary character in the worksheet did, because of the obvious dangers associated with fires in the home.**

Rusting

● Martin and his mum and dad have just moved house. The picture shows their garden. It's full of junk and looks a bit of a mess!

1. Which materials go rusty?

2. What is needed to make things go rusty?

3. Which things in the garden do you think will be rusty?

4. For each of the objects that have gone rusty, say what could have been done to them to stop them rusting.

5. If you lived near the sea, do you think that a car or bike would rust more or less quickly? Why?

Dear Helper,

We have been looking at rusting, what causes it, and how it can be prevented. Only objects made from iron and steel will rust, and they need air and water in order to start rusting. To help your child with this activity, take them out on a walk and look for things that have gone rusty. Notice what type of material the rusty objects are made from, and ask your child to think about what could have been done to stop them from going rusty.

Generating electricity

- Write a newspaper article about Michael Faraday.

- These questions will give you some ideas about what you will need to find out when you write your story.

  What did he discover?

  When did he live?

  What did his father do for a living?

  How did Michael Faraday become interested in science?

- Your article should be no longer than one side of A4 paper. You could include pictures or diagrams in your report.

- Use the space below to make notes before you write your article.

Dear Helper,

We have been looking at how electricity is made. Ask your child to tell you what is needed to make electricity before they start their research for this activity.

This activity asks your child to find out about Michael Faraday, who is associated with the discovery of electricity. Please encourage them to take notes before they begin to write their article – simply copying articles from books or the Internet should be discouraged. To make sure your child understands Faraday and his work, ask them to tell you about his life once they have written their report.

PHOTOCOPIABLE

Name:

Traffic lights

You may find coloured pencils (red, yellow and green) helpful for this activity.

● The drawing below shows a circuit to work a set of toy traffic lights. Study it carefully and then answer the questions below.

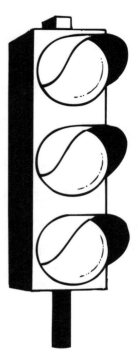

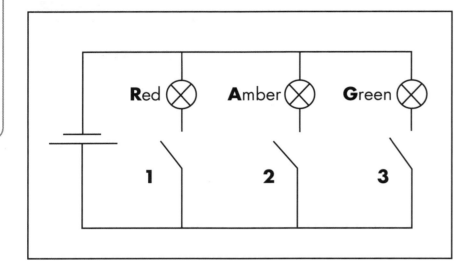

1. Which switch must be pressed to get the light or lights to switch on for the full traffic light sequence:

a Red _____

b Red and Amber _____

c Green _____

d Amber _____

e Red _____

2. Why is the brightness of the bulbs always the same, no matter how many lights are on? _____

_____

Dear Helper,

In school we have been experimenting with batteries and bulbs to make and change simple electrical circuits, like the one shown in the diagram above. It may help your child to answer question 1 if they use coloured pencils to trace the path the electricity must take to go through each bulb.

Name:

Matching up

1. Join the correct name of the electrical component to the correct symbol.

Battery

Motor

Switch

Wire

Two wires joined together

Bulb

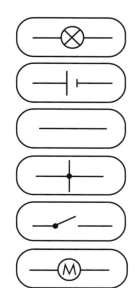

2. Draw two circuit diagrams using the correct symbols, to show:

a a motor powered by two batteries that can be switched on and off by a switch.

b two bulbs powered by two batteries. The bulbs go on and off together using a single switch.

Dear Helper,

Your child has learned in school that standard symbols are used to represent the components of an electrical circuit. This activity requires them to recall these symbols and match them up with their meanings, then to draw some simple circuits using the symbols. Watch out for these symbols in real life (for example, in the manuals for the electrical equipment in your home) and point them out to your child.

The combination is...

- The drawing below represents the wiring inside a bank. Lots of people work in the bank, but different people are allowed into different areas. Sometimes people can go into certain areas

Ms Tack
Bank Manager

Mr Swann
Assistant Manager

Toby
Cashier

alone, but in other areas they need to have another bank worker with them. When only certain people can enter a particular part of the bank, it is said to have 'restricted access'. The bank operates a security system in which each person is given a key that operates some of the bank's switches, so they can only open particular doors.

- Look at the diagrams below to find out who can work in which parts of the bank.

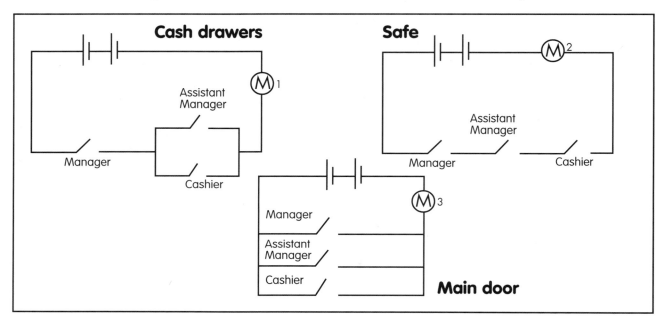

1. How many people are needed in order to open the lock controlled by motor M_1?

2. Which person must be present in order to open the cash drawers?

3. Which people, or combinations of people, can open the lock for:
   a the cash drawers? **b** the safe? **c** the main front door?

Dear Helper,

This activity requires your child to examine the way that switches can be used in real life. Electrical door combination locks are examples of where an electrical circuit must be built in the correct order so that 'throwing the switch' opens the locks that are intended. You may know of an example of this kind of system that you can show your child. When you are in town together, watch out for assistants in shops or banks using 'swipe' key cards or door codes to enter doors into restricted areas.

The safety code

● In the space below, write a set of rules for using electrical appliances and sockets.

● Write your rules in short phrases or sentences, so that people can remember them easily.

Dear Helper,

This activity connects science and literacy. It's important that your child can write in different styles, to suit the purpose of the writing and the audience who will read it. Sometimes this will mean a story style, but at other times short notes are better. Discuss with your child how they could best set out their rules, for example as *Dos and don'ts* or *You must... But you must not...*

■ SCHOLASTIC

Wired up

Length of wire/cm	Resistance/units
20	4
40	8
60	12
80	16
100	20

● This table shows the results of an experiment in which different lengths of wire were tested to see how difficult it was for electricity to flow through them. This is called **resistance**. The higher the resistance, the more difficult it is for the electricity to get through the wire. Dimmer light switches and controllers for toy racing cars work by changing the resistance in a circuit.

● Plot a line graph here to show the results.

● Complete the sentences below.

As the wire gets longer the resistance gets _____. The resistance for a 25cm wire would be _____ units. The resistance for a 50cm wire would be _____ units. The shows that when the wire is twice as long the resistance is _____.

Dear Helper,

When you increase the length of a piece of wire in an electrical circuit, it affects the way the circuit operates. All wire has *resistance*, which makes it harder for the electricity to get between the components (such as between a battery and bulb). In the figures above, we have given a value to the resistance for each piece of wire (which is measured in ohms, for example a loudspeaker labelled *8 ohms*). This activity marries up the important maths skill of graph-plotting with the common science skill of interpreting a graph to draw a conclusion.

Many hands make light work

You will need: a pencil and a helper.

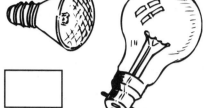

● Look around your house and find out how many different-shaped light bulbs you can see.

● How many different types of light bulb did you find? ☐

● Find a light bulb with a wire inside the glass that you can see easily. Ask your helper to remove the bulb from its holder. The light must be switched off first, and if the light bulb has been on they must take great care because the bulb could be very hot!

● Find somewhere safe to place the bulb so that you can see it clearly. Look at it together. Draw the bulb carefully in the space below. Label the parts that you have drawn.

● What kinds of material is the light bulb made from? Which parts does the electricity pass through? Fill in the details in the table.

Type of material	Does electricity go through it?

Dear Helper,

We have been looking at circuits in real life. This activity will help your child appreciate that electricity flows around all parts of their home in a circuit. If you can, show your child the main distributor point and meter where the electricity enters your home (but take care near electricity). Drawing and labelling a light bulb reinforces the idea of materials as *conductors* and *insulators* of electricity, and of using the right material for the job. To keep safe, you could secure the bulb to the table with a piece of Blu-Tack or sticky tape so it cannot roll around while they are drawing it.

PHOTOCOPIABLE

📖 SCHOLASTIC

Your next game is...

● Elizabeth is taking part in a TV game show. It's her turn to play a game for her team.

● In front of her she has found a box of materials labelled 'Magnets and bits'. Inside the box is an instruction sheet. It reads:

> **Inside this box you will find two magnets and some metal objects. You must sort them out. Here is a rhyme to help you.**
>
> 3 times we will stick, and twice we will not.
> Find the one that pushes away and you have the lot.

● The objects in the box are aluminium foil, a steel can, a small iron bar and a copper disc. There are also the two magnets, each with a North and a South Pole, of course.

1. Imagine you are on Elizabeth's team. Write down what she must do to find the answer to her problem.

_____

_____

2. What are the answers to the rhyme? Which objects are which?

3 times we will stick, _____

and twice we will not. _____

Find the one that pushes away and you have the lot.

_____

Dear Helper,

We have been looking at the force of attraction between magnets. In Year 4 (Primary 5 in Scotland), your child also looked at magnetic materials. This is an exercise to revise which materials are magnetic and to reinforce the idea of the forces between two magnets.

Newton's diary

You will need: access to an encyclopedia and research books, or multimedia CDs and the Internet.

● Sir Isaac Newton is a very famous 17th-century scientist.

● Find out more about his life and the great discoveries he made. Use the questions below to guide you through the huge amounts of information you could find. Try not to copy out large pieces of writing from a book or CD, but carefully select the information you need.

When and where was Isaac Newton born?

When did he die?

Where did Newton go to university?

What did he famously 'discover' and write about in 1665/66? (Think about apple trees!)

What did Newton find out about light?

Did Newton work with any other scientists?

Did Newton use other scientists' ideas to help him?

Newton has three famous 'Laws'. Can you find out what they are?

● People have written very large books about Isaac Newton. You don't need to write a book – not just yet anyway!

● What you are going to write now is either a short account of Newton's life that might have been published in a newspaper after his death (called an **obituary**) or an entry from Newton's diary as if he is describing a day on which he has made a wonderful new discovery.

● For the diary entry, your research should give you an idea of the year of the discovery, but you might need to make up the exact date!

Dear Helper,

Please help your child to find out as much as possible about Sir Isaac Newton (perhaps by taking them to the local library if you don't have access to enough material at home), but encourage them to write down any information they find as short points, not as long sentences in paragraphs. Please try to discourage your child from copying long sections from a book or printing pages off the Internet.

PHOTOCOPIABLE

Name: _____

Bungee!

● Before people leap off high platforms on a bungee jump, the ropes are tested for their **stretchiness** or **elasticity**.

● The graph on the right shows how one 50m rope stretches when different loads are attached to it.

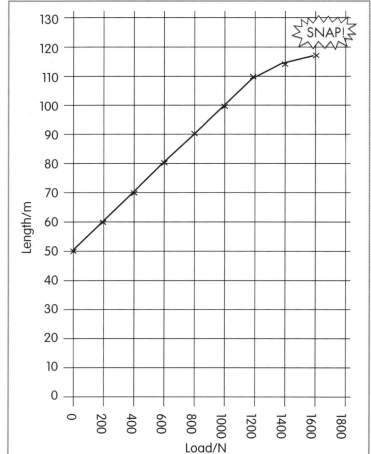

1. What is the length of the rope when there is no load attached to it? _____

2. When a load of 400N is added, how long is the rope in total? _____

3. By how much has the rope stretched when a load of 400N is added?

_____

4. What is the load on the rope when it is twice its original length? _____

5. What is the maximum length of the rope? _____

6. What is the maximum load that should be put on the rope? _____

| **Dear Helper,** |

This activity follows on from our work in class, where your child has been looking at the force needed to stretch an elastic band or spring. Being able to read information from a graph is an important skill in maths and science. If necessary, help your child to read the information off the graph in order to answer the questions.

Name:

Floating metal

You will need: a pen or pencil, a wide drinking glass or measuring jug, a small piece of tissue paper or similar, a drop or two of washing-up liquid and a needle or pin.

● This activity will show you an invisible force called **surface tension**. This force occurs on the top of a liquid, and if you are careful it is strong enough to support a piece of metal like a needle.

● Follow the instructions and diagram below.

1. Fill your container with water to about 1cm below the rim.

2. Place a square of tissue paper that is as big as your needle on the surface of the water.

3. Drop your needle on to the tissue.

4. Use the flat end of your pencil (not the writing end!) to very carefully push the tissue paper under the water all the way around the needle. It should sink to the bottom, leaving your needle on top of the water.

5. Look at the water around the needle from the side – is the surface of the water flat?

6. Put a tiny amount of washing-up liquid on your fingertip, then touch the water at the side of your container. Watch the needle carefully. What happens?

7. Wash your hands to remove the washing-up liquid and change the water before you try again.

● Draw a diagram of the 'shape' of the water around the needle.

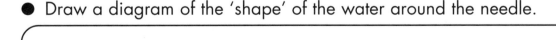

Dear Helper,

Your child has been looking at forces in the world around them. This fun activity demonstrates a force that is often ignored. *Before* they try it, talk with your child about what they think will happen when they put the washing-up liquid in the water. Remember to take care with sharp needles/pins and make sure that the washing-up liquid does not get into the eyes or mouth by mistake.

The daredevil

● Read this story, then answer the questions.

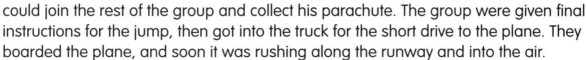

Stuart drove up to the gates of the airfield in his black sports car. The day had started cold and cloudy, but he hoped the parachute jump would still go ahead. He was directed to a large hanger by a man on the gate, so he could join the rest of the group and collect his parachute. The group were given final instructions for the jump, then got into the truck for the short drive to the plane. They boarded the plane, and soon it was rushing along the runway and into the air.

Over the jump area the group were told to get ready to jump out. Just before he jumped out of the plane Stuart said a short prayer to himself, as he looked at the ground a long long way down. He jumped out, screaming with excitement. For almost 30 seconds he fell through the air getting faster and faster. He checked his height meter and pulled the cord to release his parachute. It opened with a whoosh and Stuart slowed down very quickly. Steadily the ground came towards him as he looked for the smoke marking the landing site. 5… 4… 3… 2… 1… Stuart hit the ground with a thump.

What an experience! After a talk with the rest of his group Stuart had to go home. The clouds had gone and it was a lovely day. Stuart folded the roof back on his car and drove down the road with a huge grin on his face. What a great day!

1. What would be the difference between the drag (air resistance) affecting Stuart's sports car and the truck?

2. How would the forces acting on the plane change from going along the runway to going in the air?

3. Why did Stuart fall faster and faster when he first jumped out of the plane?

4. Why did Stuart slow down when he released his parachute?

5. What effect would wind have on the smoke coming from the landing site?

6. What advice could you give to Stuart about the forces he feels when he hits the ground?

7. How would having the roof down affect the forces Stuart would feel as his car drove along the road?

Dear Helper,

Your child may need help with reading the story, and it may be worthwhile talking through some of the events in the story before answering the questions. Ask your child to think about walking on a windy day, and look at the way flags move in the wind (some big hotels have flags outside) or take them out on a drive with the car windows or sunroof open – what forces can they feel acting on them?

The gymnastics display team

● Look at the drawings carefully and then try to answer the questions below.

1. In the table, write down where you see a force stretching or pulling and where you can see a force compressing or pushing.

Stretching/pulling forces	Compressing/pushing forces

2. What force is pulling all of the gymnasts down? _____

3. How can you tell there is a force acting on the gymnast on the bars?

_____

4. Using the ideas of forces, explain why the gymnast on the rings is not

falling to the ground. _____

Dear Helper,

This activity requires your child to examine a scene and identify how forces are affecting the way things behave or move in our surroundings. Extend this activity by discussing situations that you may see on television or when you are out, such as circus performers or monkeys in a zoo, or even the forces you experience if you are on the swings in a local park. Ask your child to tell you what forces they can see in these situations and how they are acting on the people or objects involved.

Name:

Staying in the shade

● This picture shows a park scene. Look at the position of the Sun and the objects in the park carefully, then try to answer the questions below.

A   B   C   D

1. The Sun is shining brightly. What will the temperature be like?

_____

2. Why are some people sitting under the tree? _____

3. Which letter on the diagram – **A**, **B**, **C** or **D** – will the shadow from the top of the tree touch? _____

4. Why does the shadow appear? _____

5. What would happen to the shadow if a cloud passed in front of the Sun?

_____

Dear Helper,

This activity revises how shadows are formed in our environment. Ask your child to explain – remember that any light source passing an object creates a shadow the same shape as the object and directly opposite the source of light. On a sunny day, take your child out to look for the features of the picture above.
⚠ **It is essential that you DO NOT allow you child to look directly at the Sun when comparing the position of the shadow to the position of the Sun.**

True or false?

- Read through these statements about how we see objects in our environment.

- Think about them carefully and decide if they are true or false. Tick the box if the statement is true.

☐ A set of Christmas tree lights is a light source.

☐ Mirrors reflect light.

☐ The Moon is not a light source.

☐ We see objects because light enters our eyes.

☐ The Sun produces its own light.

☐ At night we can wear special clothes that reflect light to make us easier to see.

☐ All the stars in the sky are light sources.

☐ We see our family because light from a light sources scatters off them into our eyes.

☐ We see the Moon because the light reflects off the Sun.

☐ Objects that produce their own light are called light sources.

Dear Helper,

This activity is designed to reinforce your child's learning about the way light travels around our environment, allowing us to see things. It also addresses some common misconceptions about light that can be made by children and adults alike – so be careful!

Seeing things differently

You will need: a pencil, a pen (optional) and a large spoon (remember to ask before you borrow one!).

● Go around your home with your helper, and try to find reflective surfaces. You will probably find lots of mirrors that reflect, but can you find anything else?

● Make a list of the things you find.

● Did you include a spoon in your list? Borrow a large spoon and look at yourself in it. Try moving the spoon away from you and close up, and looking in both sides of the spoon.

● Draw some pictures of what you look like in the spoon. Remember to label the way you held the spoon for the image you saw.

Dear Helper,

Reflective surfaces are all around us, and there are many more than just mirrors even in the home. This activity explores how a mirror-like curved surface gives a very different view of the world around us. You may be able to find other curved mirrors for your child to look at – try the reflectors in torches or headlamps on cars, the mirrors on your car or the security mirrors in a shop. As with all glass, take great care to avoid breakages and cut fingers!

What a view!

You will need: a pen or pencil, and a picture of a landscape (a photo or picture from a magazine of buildings, a town or the countryside – it could be anywhere).

● Examine your picture carefully. Look for sources of light, shadows and reflections. See how they work together to make the place what it is.

● In the space below, describe the light in your picture. Look particularly at how the shapes are made by the light and shadows. Describe how the picture makes you feel. Attach your chosen picture to this sheet if you can.

Dear Helper,

This homework helps to develop your child's literacy skills using images that include elements of science. Help your child to choose a picture of a place with strong areas of light and shadow, for example light passing through a wood or past some clouds where beams of light can be seen. Talk with them about the picture before they write their thoughts down – try using words like *bright, dark, warm, cold, shiny* or *dull*.

Name:

On the way

You will need: some coloured pencils.

I was woken up this morning by my mum knocking on my bedroom door. I could hear the cars in the street and the sound of somebody running the shower. I got out of bed and went down for breakfast. When I finished my breakfast, Mum wasn't happy as I bashed the spoon into the bowl.

I went upstairs to wash and get dressed. Mum called from downstairs – we were going to be late for school if I didn't hurry up. We went out to the car. The door slammed shut. The engine started and off we went.

On the way to school an angry driver was beeping his horn and shouting at a truck driver who had stopped his truck in the way.

Mum dropped me off at school; I could hear all my friends shouting in the playground. The teacher rang the bell for the start of school and we all went inside. As we went in the hinges on the old door squeaked as it swung back and forth. In class, the teacher called out our names on the register and we answered, "Yes, Mr Buckingham." This was the start of another day at school.

1. In the story, underline the occasions when a person makes a sound with a coloured pencil.

2. With a different-coloured pencil, underline where an object makes a sound.

3. What do we call an object that makes sound?

4. How does an object making a sound move?

Dear Helper,

This activity requires your child to identify *sources* of sound, and to describe how sound travels from a source to our ears. You could try this activity on any journey or during a quiet moment – rather than 'I-spy', play 'Ear 'ere' to identify what sounds you can hear and their sources.

The speed of sound

● The data below shows how fast sound can travel in different materials. The list contains solids, liquids and gases.

Material	Speed of sound (m/s)
Air	330
Water	1500
Glass	3800
Wood	3500
Iron	5000
Brick	5000

● Draw a bar chart to show this data.

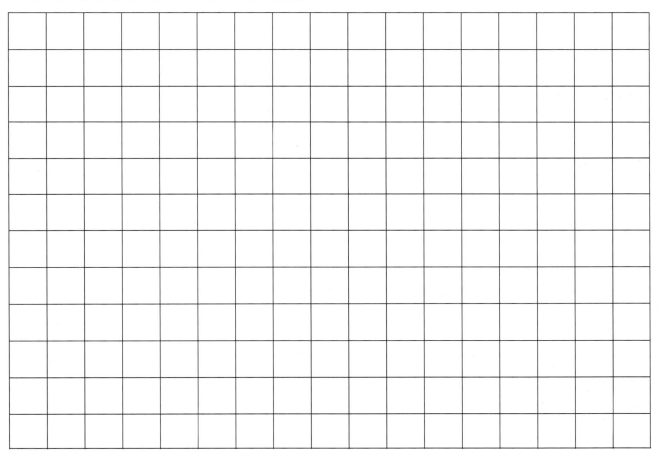

Dear Helper,

This activity uses scientific data to give your child practice in an important maths skill: converting data from a table to a graph. Each material should have a separate bar on the chart. Discuss a sensible vertical scale up the side of the graph. Ask your child: *What should 1cm on the graph represent? 10m/s? 100m/s? 1000m/s? Let your child decide!*

Name:

The Sun, Moon and the Earth

1. Here is a picture of Earth as seen from space.

© Digital Vision Ltd

Draw an arrow on the picture to show from which direction the Sun is shining on to the Earth.

2. Describe two ways in which the Earth moves during a year.

_____

3. The Moon does not give off its own light. Complete the following sentence.

The Moon shines with light _____ from the _____.

4. Seen from the Earth, the Moon seems to change shape. How long does it take for the Moon to go from full Moon to full Moon? _____

_____

5. Complete the following sentences.

The Earth spins on its axis once every _____ hours.

The half of the Earth facing away from the Sun is in _____.

It takes about 365 days for the Earth to orbit the _____.

Dear Helper,

We have been looking at the size of the Sun, Moon and the Earth in relation to one another, as well as the relative distance between them and the movements of the Moon and the Earth. This activity is to help your child revise the facts covered in the lesson.

An eclipse of the Sun

● Complete this passage with words from the box below. The words can be used once, more than once or not at all:

Earth	Moon
Sun	eclipse
umbra	further
bigger	solar

© NASA

An eclipse of the Sun (called a _____ eclipse) happens when the Earth, Moon and Sun line up in such a way that the Moon blocks the Sun's light from the _____. Such an eclipse only occurs when the _____ lies directly between the Earth and the Sun. The Moon's shadow, or _____, only covers a small area of the Earth's surface. Anyone standing in this region will see a total solar eclipse.

Eclipses occur because the Sun and the Moon appear to be the same size in the Earth's sky. In reality, the Sun is 400 times _____ than the Moon; but because the Sun is 400 times further away, it appears the same size as the Moon.

If the Moon were _____ away from the Earth, it would not cast such a large shadow on the Earth when it passed between the Earth and the Sun.

Dear Helper,

You probably remember the solar eclipse that occurred in the summer of 1999. We have been learning about what causes a solar eclipse. This activity revises the facts covered in the lesson. If necessary, read it through with your child, and emphasise the need for careful reading and correct spellings.

Name:

The life and work of Galileo Galilei

● Galileo Galilei lived a long time ago. He is famous for his scientific discoveries. Find out as much as you can about him, then write a short report about his life.

● The following points might help you plan your project:

When did he live?

Where did he live?

What scientific object did he make? Was the idea totally his?

What did he observe and study?

What disability did he suffer from later on in his life? What might have caused it?

Why was he imprisoned?

● Make notes below. Remember to record the titles of the books you look in and the websites you visit.

Dear Helper,

Your child has been looking at the main events in Galileo's life. Here, they will need to do some research in order to write a report about his scientific discoveries. They will have been told which events to focus on and which events to avoid. Please encourage note-taking, as opposed to simply printing out pages from the Internet or a CD-ROM. To make sure that your child understands what they have written, it would be a good idea to ask them to tell you about Galileo and his life when they have written their report.

Name:

The Solar System

1. Complete the following sentence. Tick one box.

The Voyager space probe sent back pictures of Jupiter and Saturn to Earth. The pictures showed us more about:

the Milky Way ☐   the Solar System ☐   galaxies ☐

the Universe ☐   the Moon ☐

2. Put these objects in order of size.

planet	Solar System	Universe	galaxy

Smallest _____

_____

_____

Largest _____

3. The Earth and other planets orbit the Sun. Label the planets shown on the diagram below:

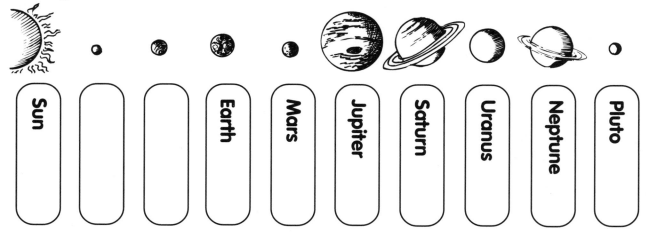

| Sun | | | Earth | Mars | Jupiter | Saturn | Uranus | Neptune | Pluto |

4. Can you remember the order of the planets from the Sun? Make up a rhyme to help you remember the order and write it in the space below.

Dear Helper,

We have been learning about our Solar System and what it consists of. If your child is stuck, the order of the planets is shown in question 3, and you could suggest *My Very Easy Method Just Speeds Up Naming Planets* as a rhyme to remember the order of the planets.

Name:

How far from the Sun?

● Some children are making a bar chart showing the distance of the planets from the Sun.

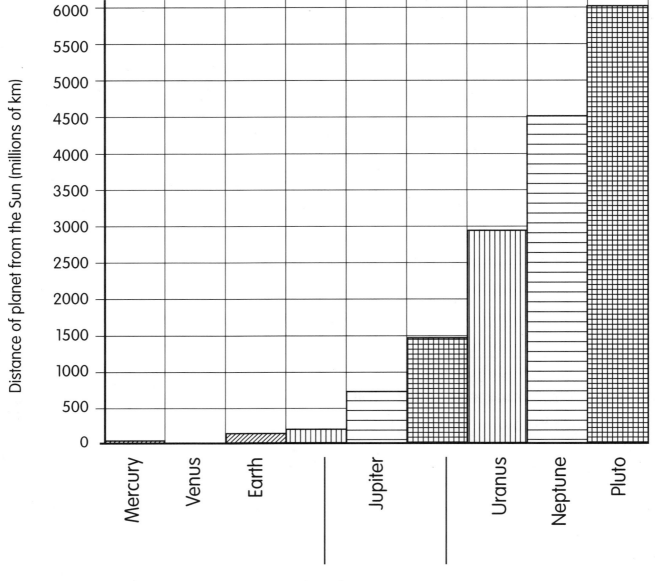

1. Complete the missing labels on the chart.

2. The bar showing how far Venus is from the Sun is missing from the chart. Draw a bar about the correct height on the children's chart.

3. How far is Earth from the Sun? _____

4. Which planet is approximately 1500000km from the Sun? _____

5. How much further is Uranus from the Sun than Jupiter?_____

Dear Helper,

We have made a model of the Solar System in class, using a scale that shows the relative distance between each of the planets and the Sun. The graph on this page shows this information, and will help your child practise reading and interpreting data. Help your child read the information from the graph, if necessary.